# Field Theories in Condensed Matter Physics
## A WORKSHOP

# Field Theories in Condensed Matter Physics

## A WORKSHOP

Edited by Zlatko Tesanovic

**ADDISON-WESLEY PUBLISHING COMPANY**
*The Advanced Book Program*
Redwood City, California • Menlo Park, California • Reading, Massachusetts
New York • Don Mills, Ontario • Wokingham, United Kingdom • Amsterdam
Bonn • Sydney • Singapore • Tokyo • Madrid • San Juan

Publisher: Allan M. Wylde
Production Manager: Jan V. Benes
Promotion Manager: Laura Likely
Cover Design: Irene Imfeld Graphic Design

**Library of Congress Cataloging-in-Publication Data**

Field theories in condensed matter physics/Zlatko Tesanovic
  [editor].
    p. cm.
  Includes index.
  1. Condensed matter. 2. Statistical physics. 3. Field theory
(Physics) I. Tesanovic, Zlatko.
QC173.4.C65F54   1990     530.4'1—dc19     90-363
ISBN 0-201-50391-3

This book was prepared by the author, using the TEX typesetting language.

Copyright © 1990 by Addison-Wesley Publishing Company, The Advanced Book Program,
350 Bridge Parkway, Redwood City, CA 94065.

All rights reserved. No part of this publication may be reproduced, stored in a retrieval system, or transmitted in any form or by any means, electronic, mechanical, photocopying, recording, or otherwise, without the prior permission of Addison-Wesley Publishing Company.

ABCDEFGHIJ-MA-943210

# Foreword

The Workshop on *Field Theories in Condensed Matter Physics* took place at the Theoretical Interdisciplinary Physics-Astrophysics Center (TIPAC) of the Johns Hopkins University from November 3 to 5, 1988. The purpose of the Workshop was to bring together a selected group of researchers actively working on various field theoretical aspects of condensed matter physics problems.

In recent years we have witnessed a great surge in the use of advanced field theory methods to tackle some of the most interesting issues in condensed matter physics. Rather than playing with highly abstract mathematical models or providing an elegant and (sometimes) rigorous packaging for already prefabricated physical ideas, many field theorists found themselves at the front lines of physics in the making, often being the first to offer a physical insight into a newly discovered phenomenon. Many others became field theorists. Initially trained in a more traditional way, they were driven to field theory by its inherent beauty but even more by their own desperation, for no other approaches worked. Thus, slave-bosons, nonlinear sigma models, anyons and instantons became an integral part of the vocabulary at the forefront of research in condensed matter physics. These Proceedings represent only a small cross-section of the overall activity in field theory methods. But it is a carefully chosen cross-section. Most of the major fields are represented: high temperature superconductivity, low-dimensional magnetism, quantized Hall effect, localization, etc., and the contributors have played a major role in the development of their subjects. Bearing in mind that their topics often place a heavy (although ultimately rewarding!) burden of an increased formal sophistication on many a practicing theorist (let alone graduate students), they have attempted, to the best of their abilities and against heavy demands on their time, to be pedagogical and to provide references for further study. In rapidly evolving research, where the only way to learn is by plowing through current physics periodicals, this is not an insignificant detail.

This Workshop has been made possible by the financial support of the Faculty of Arts and Sciences at The Johns Hopkins University. The job of making sure everything ran smoothly was handled very successfully by Nicole Carr, Lenore Danielson, and Karen Diener, as well as by our graduate students and postdoctoral fellows in theoretical condensed matter physics, too numerous to mention by name.

We wish to thank Allan M. Wylde, Jan Beneš and the staff of Addison-Wesley Publishing Company for their understanding and excellent work in producing these Proceedings.

Zlatko Tešanović for<br>
the Organizing Committee

# Contents

1     Foreword     v

2     Schwinger Boson Mean Field Theory of the Quantum Heisenberg Model
         Assa Auerbach and Daniel P. Arovas     1

3     The Theory of the Collective Excitations in the A-Phase of $He^3$
         P. N. Brusov     17

4     Induced Quantum Numbers and Surface Currents in Some 2 + 1 Dimensional Models
         Yi-Hong Chen and Frank Wilczek     39

5     The Superfluid Density Tensor in Unconventional Superconductors
         C. H. Choi and Paul Muzikar     51

6     Renormalized Fermi Liquid Theory for Disordered Electron Systems and the Metal-Insulator Transition
         C. Di Castro     59

7     The Spectrum of Short-Range Resonating Valence Bond Theories
         Eduardo Fradkin     73

8     Antiferromagnetic Phase Instability in the Two-Impurity Kondo Problem
         B. A. Jones     87

9     Field Theories for Elasticity of Tethered Networks
         Mehran Kardar     105

10     Statistics of Holons in the Quantum Hard-Core Dimer Gas
         Steven Kivelson     113

11     Conductivity and Tunnelling Density of States Exponents at the Metal Insulator Transition with Strong Spin Orbit Scattering
         Gabriel Kotliar and Sandro Sorella     125

12     Coulomb Potential in Layered Metals, Bloch States, Wannier Functions and High-$T_c$ Superconductivity
         Daniel C. Mattis     139

## Field Theories

13 The Two Kondo Impurity Problem: A Large $N$ Biased Review
    A. J. Millis, B. G. Kotliar, and B. A. Jones     159

14 A New Crossover Behavior of Type II Superconductors in Strong Magnetic Fields
    M. Rasolt and Z. Tesanovic     167

15 Valence-Bond and Spin-Peierls Ground States of Low-Dimensional Quantum Antiferromagnets
    N. Read and Subir Sachdev     179

16 The Landau-Ginzburg Theory for the Fractional Quantum Hall Effect
    S. C. Zhang, T. H. Hansson, and S. Kivelson     189

17 One-Dimensional Magnets, Wess-Zumino Models and Conformal Invariance
    Timothy Ziman     199

**Assa Auerbach**
Physics Department
Boston University
Boston, MA 02215
and
**Daniel P. Arovas**
Physics Department
University of California at San Diego
La Jolla, CA 92093

# Schwinger Boson Mean Field Theory of the Quantum Heisenberg Model

We review the Schwinger boson mean field ("large $N$") theory for the Heisenberg ferromagnet and antiferromagnet. Comparison with exact solutions, semiclassical field theories and numerical results suggests that this approximation is valid and quantitatively successful in the regime where topological Berry-phase terms are unimportant.

## I. INTRODUCTION

Much progress has been recently made in understanding the quantum Heisenberg model (QHM). Since in one and two dimensions, quantum and thermal fluctuations combine to destroy long range order,[1] the use of naïve spin wave theory[2] is strictly limited to zero temperature and to ground states with broken rotational symmetry. For the antiferromagnetic model in one dimension, it can be that the *ground state* has no long range order, as shown by Bethe's solution[3] of the $S = 1/2$ chain, and Haldane's mapping of the large $S$ QHM onto the nonlinear sigma model (NLSM) in $(1 + 1)$ dimensions[4] ($S$ is the length of the spin). The field theoretical approach includes a topological $\theta$ term, which differentiates between the integer and half-odd-integer $S$ ground states,[5] in accordance with the Lieb-Shultz-Mattis theorem.[6] This approach was recently extended to the two dimensional model, where an ordered ground state exists above a critical value of the spin. In the disordered phase (small $S$ and/or large frustration), interference from topological terms was predicted[7] to depend on the value of $2S$ mod 4.

Here we shall review an alternate asymptotic approach[8,9] based on the functional integral steepest descents approximation to the QHM. The resulting Schwinger boson mean field theory (SB-MFT) is the large $N$ theory of a particular generalization of the QHM to SU($N$) generators replacing the usual SU(2) spin operators. Comparisons of our results to exact and numerical calculations for the physical $N = 2$ model reveal a surprising success of the low order approximation.

Although the SB-MFT reproduces the effects given by the NLSM field theory even for small values of $S$, the aforementioned topological terms are missing. During this workshop, Nick Read[10] has provided an illuminating discussion of how these terms should be properly incorporated in the functional integral.

The generalized SU($N$) Hamiltonian is given by

$$H = \frac{1}{N} \sum_{\substack{(ij) \\ \alpha,\beta}} S^\alpha_\beta(i) S^\beta_\alpha(j) = \frac{1}{2} \tilde{S}^2$$

$$= -\frac{1}{N} \sum_{(ij)} :\mathcal{A}^\dagger_{ij} \mathcal{A}_{ij}: \tag{1.1}$$

$$\mathcal{A}_{ij} \equiv \sum_\alpha b^\dagger_{\alpha i} b_{\alpha j} \quad \text{ferromagnet,}$$

$$\mathcal{A}_{ij} \equiv \sum_\alpha b_{\alpha i} b_{\alpha j} \quad \text{antiferromagnet,}$$

where in each bond $\langle ij \rangle$ the site $j$ is taken to be in the second sublattice. Here the generalized spin operators are defined as:

$$S^\alpha_\beta(i) \equiv b^\dagger_{\alpha i} b_{\beta i}, \tag{1.2}$$

which satisfy the algebra

$$[S^\alpha_\beta(i), S^\rho_\sigma(j)] = \delta^\rho_\beta \delta^j_i S^\alpha_\sigma(i) - \delta^\alpha_\sigma \delta^i_j S^\rho_\beta(i) \tag{1.3}$$

and are subject to the constraint $\sum^N_{\alpha=1} S^\alpha_\alpha(i) = NS$. $S$ must be an integer multiple of $1/N$. For the ferromagnet, $\tilde{S}$ is defined by Eq. (1.2), while for the antiferromagnet,

$$S^\alpha_\beta(i) \equiv -b^\dagger_{\beta i} b_{\alpha i}, \tag{1.3'}$$

We shall review our theory for the case of the ferromagnet and antiferromagnet separately.

## II. THE FERROMAGNETS

The ferromagnetic partition function is given by[8]

$$Z_F = \int \mathcal{D}[b, \bar{b}; Q, \bar{Q}; \lambda] \exp(-\mathcal{I}_F[b, \bar{b}; Q, \bar{Q}; \lambda])$$

$$\mathcal{I}_F = \int_0^\beta d\tau \left\{ \frac{1}{2} \sum_\alpha (\bar{b}_{\alpha i} \dot{b}_{\alpha i} - \bar{b}_{\alpha i} b_{\alpha i}) + N \sum_{\langle ij \rangle} \bar{Q}_{ij} Q_{ij} \right. \quad (2.1)$$

$$\left. + \sum_{\langle ij \rangle \atop \alpha} (\bar{Q}_{ij} \bar{b}_{\alpha i} b_{\alpha j} + Q_{ij} b_{\alpha i} \bar{b}_{\alpha j}) + \sum_{i,\alpha} \lambda_i (\bar{b}_{\alpha i} b_{\alpha i} - S) \right\}.$$

Making the static assumption

$$Q_{ij}^{MF}(\tau) = Q$$
$$\lambda_i^{MF}(\tau) = \lambda, \quad (2.2)$$

the Schwinger bosons can be integrated out explicitly, resulting in a free enery of

$$F^{MF} = \frac{1}{2} z Q^2 - S\lambda + \frac{1}{\beta} \int \frac{d^d k}{(2\pi)^d} \ln(1 - e^{-\beta \omega_k}), \quad (2.3)$$

where $z$ is the lattice coordination number, $d$ is the number of spatial dimensions, $\mathcal{N}$ is the total number of sites in the lattice, and the integral is performed over the first Brillouin zone. The dispersion $\omega_k$ is defined by

$$\mu \equiv \lambda - zQ$$
$$\epsilon_k \equiv \frac{1}{2} \sum_\delta (1 - e^{ik \cdot \delta}) \quad (2.4)$$
$$\omega_k \equiv \mu + zQ\epsilon_k.$$

In Fig. 1, we plot $\omega_k$ in one dimension. The saddle point equations $\delta F/\delta Q = 0$ and $\delta F/\delta \lambda = 0$ are

$$S = \int \frac{d^k k}{(2\pi)^d} n_k \quad (2.5)$$

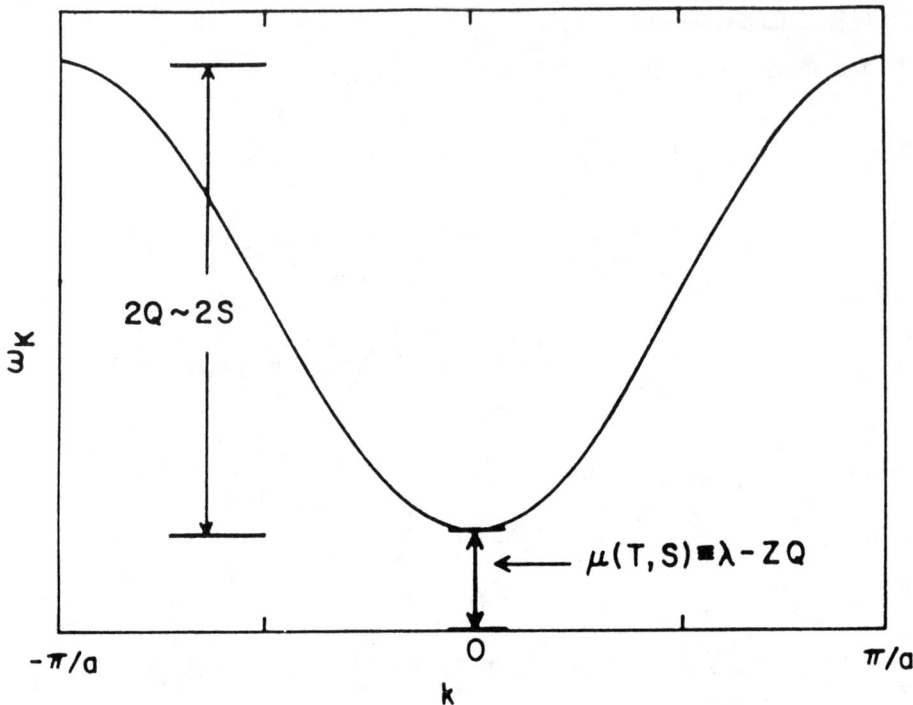

FIGURE 1 Schwinger boson excitation spectrum in the mean field theory for the ferromagnetic chain. See Eq. (2.4).

$$Q = S - \int \frac{d^d k}{(2\pi)^d} \epsilon_k n_k \qquad (2.5)$$

with $n_k = (e^{\beta \omega_k} - 1)^{-1}$. Thus, we obtain a free energy per spin of

$$F^{MF} = -\frac{1}{2} z S^2 + \frac{1}{2} z(Q - S)^2 - S\mu - \frac{1}{\beta} \int \frac{d^d k}{(2\pi)^d} \ln(1 + n_k) \qquad (2.6)$$

Upon addition of the rreference energy (see Eq. (1.1)) $+ \frac{1}{2} z S^2$, and taking $N = 2$, the first term gives the classical ferromagnetic ground state energy per spin, $E_0^{cl} = -\frac{1}{2} z^2$. We note that the remaining contribution is precisely *twice* Takahashi's result for $(F - E_0^{cl})$.[11] This factor of two is easily seen to be an artifact of the static constraint and is a generic consequence of approximations of this sort. The SU(N) theory is defined in terms of N bosons and 1 constraint (per site). Uniformizing the field λ amounts to ignoring the nonzero wavelength components of the constraint field, enforcing the local restriction $\Sigma_\alpha b_\alpha^\dagger b_\alpha = NS$ only on average, cf Eq. (2.5a). Thus, at the mean field level, the number of independent degrees of freedom is overcounted by a factor $g = N/(N-1)$. This is partially corrected by the $\mathcal{O}(1/N)$ contribution $F^{(1/N)}$ arising from integration over the gaussian fluctuations of the constraint field, as was demonstrated in Ref. [8].

The mean field equations which determine $Q(T, S)$ and $\mu(T, S)$ are identical to those of Takahashi,[12] and we have independently verified his solutions (details may be found in Ref. [11]). From Eq. (3.6), we obtain for the one-dimensional chain

$$(F^{MF} - E_0^{MF})_{\text{chain}} = T\left\{ -\frac{\zeta(3/2)}{\sqrt{2\pi}}\left(\frac{T}{2S}\right)^{1/2} + \frac{1}{2S}\left(\frac{T}{2S}\right) + \mathcal{O}(T^{3/2})\right\}. \quad (2.7)$$

where $E_0^{MF} \equiv F^{MF}(T = 0)$, which is an expansion in the quantity $T/S$, assumed here to be small. The calculation of $F^{(1/N)}$ was carried out[8] in appendix B, where it was found (Eq. (B.9)):

$$F^{(1/N)} = -\frac{1}{N}\frac{\zeta(3/2)}{2\sqrt{\pi S}} T^{3/1} + \mathcal{O}(T^{5/2}). \quad (2.8)$$

Combining Eqs. (3.7, 3.8) and setting $N = 2$ yields, to $\mathcal{O}(1/N)$,

$$(F - E_0^{MF}) = 2(F^{MF} - E_0^{MF} - F^{(1/N)})$$

$$= -\sqrt{\frac{2}{\pi}}\zeta(3/2)\left(\frac{T}{2S}\right)^{3/2} + \frac{T^2}{2S^2} + \mathcal{O}(T^{5/2}). \quad (2.9)$$

Comparing our expression with that of Takahashi, we see that the $\mathcal{O}(1/N)$ corrections have brought our $\mathcal{O}(T^{3/2})$ term in line with his, but that our coefficient of the $\mathcal{O}(T^2)$ term remains a factor of 2 too large. The Takahashi result is in remarkable agreement with thermodynamic Bethe *Ansatz* results for $S = \frac{1}{2}$. One unfortunate aspect of Takahashi's variational density matrix is that it is not rotationally invariant, and therefore the longitudinal and transverse susceptibilities is his model will be unequal. Takahashi calculates the static susceptibility

$$\chi = g^2\beta \frac{1}{N}\sum_{i,j}\langle S_i^z S_j^z \rangle \quad (2.10)$$

by performing a rotational average of $\langle (S_i \cdot \hat{n})(S_j \cdot \hat{n})\rangle$ and finds the corresponding result to be in good agreement with known $S = \frac{1}{2}$ results. That this rotational averaging produces the 'correct' result is interesting, although we wish to emphasize that Takahashi's underlying theory is not rotationally invariant. Our model preserves rotational invariance, and we find

$$\begin{aligned}\chi_{\text{chain}} &= \frac{1}{2}g^2 \int_{-\pi}^{\pi} \frac{dk}{2\pi} n_k(1+n_k) \\ &= g^2 S^4 T^{-2} \left\{ 1 - \frac{3}{S} \frac{\zeta(3/2)}{\sqrt{2\pi}} \left(\frac{T}{2S}\right)^{1/2} + \mathcal{O}(T) \right\} \end{aligned} \quad (2.11)$$

which is 3/2 as great as Takahashi's result. For the two-dimensional square lattice, we find

$$\begin{aligned}(F^{MF} - E_0^{MF})_{\text{sq}} &= -\frac{1}{2} T^2 \left\{ \frac{\zeta(2)}{2\pi S} + \frac{\zeta(3)}{8\pi S} \left(\frac{T}{2S}\right) + \mathcal{O}(T^2) \right\} \\ \chi_{\text{sq}} &= \frac{g^2}{2\pi S} \exp\left(\frac{4\pi S^2}{T}\right) + \mathcal{O}(Te^{4\pi S^2/T}). \end{aligned} \quad (2.12)$$

The spin-spin correlation function is

$$\begin{aligned}\langle \mathbf{S}_0 \cdot \mathbf{S}_R \rangle &= \frac{3}{2} |f(\mathbf{R})|^2 \\ f(\mathbf{R}) &\equiv \int \frac{d^d k}{(2\pi)^d} e^{i\mathbf{k}\cdot\mathbf{R}} n^k. \end{aligned} \quad (2.13)$$

At long distances, one is concerned with the small $k$ behavior of the occupation function $n_k$, and we obtain the following asymptotic expressions:

$$\langle \mathbf{S}_0 \cdot \mathbf{S}_R \rangle \sim \frac{3}{2} S^2 e^{-R/\xi}$$

$$\xi \sim S^2/T; \quad (d=1) \quad (2.14)$$

$$\langle \mathbf{S}_0 \cdot \mathbf{S}_R \rangle \sim \frac{3T^2}{8\pi S^2} \frac{e^{-R/\xi}}{(R/\xi)}$$

$$\xi \sim \sqrt{S/T} \exp(2\pi S^2/T); \quad (d=2, \text{square}).$$

As discussed in Ref. [11], Eq. (2.14) differs only in its prefactor from the Ornstein-Zernike correlation function expected for the two-dimensional *classical* Heisenberg model.[13]

## III. THE ANTIFERROMAGNETS

The bosonic partition function for the spin-$S$ Heisenberg antiferromagnet is given by:

$$Z_A = \int \mathcal{D}[b, \bar{b}; Q, \bar{Q}; \lambda] \exp(-\mathcal{T}_A[b, \bar{b}; Q, \bar{Q}; \lambda])$$

$$\mathcal{T}_A = \int_0^\beta d\tau \left\{ \frac{1}{2} \sum_{i,\alpha} (\bar{b}_{\alpha i} \dot{b}_{\alpha i} - \bar{b}_{\alpha i} b_{\alpha i}) + N \sum_{\langle ij \rangle} \bar{Q}_{ij} Q_{ij} \right. \quad (3.1)$$

$$\left. + \sum_{\langle ij \rangle \atop \alpha} (\bar{Q}_{ij} b_{\alpha i} b_{\alpha j} + Q_{ij} \bar{b}_{\alpha i} \bar{b}_{\alpha j}) + \sum_{i,\alpha} \lambda_i (\bar{b}_{\alpha i} b_{\alpha i} - S) \right\}.$$

The mean field theory amounts to a steepest descents approximation, where $Q$ and $\lambda$ acquire static uniform values, that are determined by extremizing the free enery. The mean field (MF) Hamiltonian is given by:

$$H^{MF} = \frac{1}{2} \mathcal{N}^2 - \mathcal{N} N S \lambda + \frac{1}{2} \sum_{k,\alpha} [\lambda (b^\dagger_{k\alpha} b_{k\alpha} + b^\dagger_{-k\alpha} b_{-k\alpha})$$

$$+ zQ(\bar{\gamma}_k b_{k\alpha} b_{-k\alpha} + \gamma_k b^\dagger_{k\alpha} b^\dagger_{-k\alpha}], \quad (3.2)$$

with

$$\gamma_k \equiv \frac{1}{z} \sum_\delta t_\delta e^{-ik\cdot\delta}. \quad (3.3)$$

It can easily be verified that the Hamiltonian does not break rotational symmetry. $H^{MF}$ is readily diagonalized by the quasiparticle operators: $\alpha_{k\alpha} = \cosh\theta_k b_{k\alpha} + \sinh\theta_k b^\dagger_{-k\alpha}$. Here, $\tanh(2\theta_k) = -zQ\gamma_k/\lambda$. Thus, the mean field free energy is given by

$$\frac{1}{N} F^{MF} = \frac{1}{2} zQ^2 - \frac{1}{2}(2S+1)\lambda + \frac{1}{\beta} \int \frac{d^d k}{(2\pi)^d} \ln\left(2 \sinh \frac{1}{2} \beta \omega_k \right)$$

$$\omega_k = \sqrt{\lambda^2 - z^2 Q^2 |\gamma_k|^2}. \quad (3.4)$$

and the steepest descents equations are

$$\frac{1}{N} \frac{dF^{MF}}{d\lambda} = \int \frac{d^d k}{(2\pi)^d} \cosh(2\theta_k) \left(n_k + \frac{1}{2}\right) - \left(S + \frac{1}{2}\right) = 0 \quad (3.5)$$

$$\frac{1}{N}\frac{dF^{MF}}{dQ} = -z\int\frac{d^d k}{(2\pi)^d}\gamma_k \sinh(2\theta_k)\left(n_k + \frac{1}{2}\right) + zQ/J = 0. \tag{3.6}$$

Here, $n_k$ is the Bose occupation $[\exp(\omega_k/T) - 1]^{-1}$. We have shown[8] that the stable mean field solution has $t_\delta = 1$. The structure factor for the SU(2) model (two Schwinger boson flavors) is defined as: $S \equiv \langle S^z(\mathbf{q}, \omega)S^z(-\mathbf{q}, -\omega)\rangle$, and in the mean field level is given by:

$$S^{MF}(\mathbf{q}, iq_n) = \frac{1}{2}\int_0^\beta d\tau e^{iq_n\alpha}\sum_{kk'}\langle T\{b^\dagger_{k1}(\tau)b_{k+q1}(\tau)b^\dagger_{k'+q1}(0)b_{k'1}(0)\}\rangle, \tag{3.7}$$

where $q_n$ is a Matsubara frequency. Here it is convenient to measure the reduced momentum with respect to the antiferromagnetic vector $\pi$; i.e. $\tilde{\mathbf{q}} \equiv \mathbf{q} - \pi$. In (3.7) we have exploited the decoupling of the different Schwinger bosons at the mean field level.

It is convenient to parametrize the dispersion $\omega_k$ in terms of the spin wave velocity $c = \sqrt{8}Q$, and a parameter $\kappa$ such that $c\kappa = \sqrt{8(\lambda^2 - (4Q)^2)}$. The dispersion is then given by $\omega_k = c\sqrt{(\kappa/2)^2 + 2(1 - \gamma_k^2)}$. We note that $\kappa/2$ serves as a cut-off in the momenta integrations in Eqs. (3.5, 3.6). Our spin waves are therefore "massive" when $\kappa$ is finite. In Fig. 2, we plot $\omega_k$ in one dimension. It is possible to write down the (unrenormalized) projected variational ground state corresponding to the mean field approximation:

$$|\Psi_0\rangle = \mathcal{P}\exp\left(\sum_{m,n,\alpha} u(\mathbf{n})^\dagger_{m,\alpha}b^\dagger_{m+1,\alpha}\right)|0\rangle \tag{3.8}$$

where

$$u(\mathbf{n}) = \int\frac{d^d k}{(2\pi)^d}\tanh\theta_k e^{ik\cdot n}$$
$$\tanh\theta_k = (\sqrt{1 - \eta^2\gamma_k^2} - 1)/\eta\gamma_k. \tag{3.9}$$

Here, $\mathcal{P}$ is the projector of the constraint, which eliminates all components with the wrong number of Schwinger bosons at any site. It is easy to see that $u(\mathbf{n})$ vanishes unless $e^{i\pi\cdot n} = -1$, i.e. $\mathbf{n}$ connects one sublattice to the other. Now if we consider the $N = 2$ model and reverse our sublattice rotation, we find

$$|\Psi\rangle = \mathcal{P}\exp\left(2\sum_{\substack{m\in\mathcal{A}\\m+n\in\mathcal{B}}}u(\mathbf{n})(a^\dagger_m b^\dagger_{m+n} - b^\dagger_m a^\dagger_{m+n})\right)|0\rangle, \tag{3.10}$$

where the sum on $\mathbf{m}$ is over the $\mathcal{A}$ sublattice only. Thus $|\Psi\rangle$ is a sum of states each of which is constructed by successive application of the (rotationally invar-

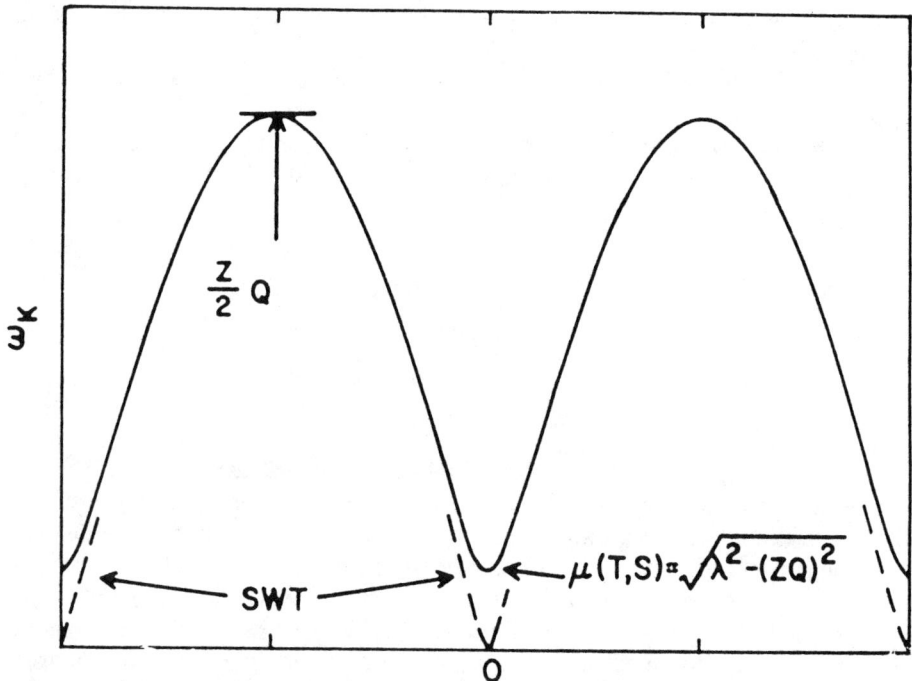

FIGURE 2 Schwinger boson excitation spectrum in the mean field theory for the antiferromagnetic chain. See Eq. (2.4). Gapless spin waves of the naïve spin wave theory (SWT) are marked with dashed lines.

iant) composite operator $\mathcal{C}^\dagger_{m,m+n} = (a^\dagger_m b^\dagger_{m+n} - b^\dagger_m a^\dagger_{m+n})$ between sites on alternate sublattices, the amplitude for each configuration in the sum given by the product of the associated $u(\mathbf{n})$'s. The presence of $\mathcal{P}$ means that only configurations where $2S$ such "bonds" emanate from *each* site are present. If we consider the $S = 1$ chain and take $\eta \to 0$, ignoring for the moment the fact that $\eta$ is a function of $S$ through Eqs. (3.5, 3.6) (one could imagine adding terms to $H$ which would lead to an $S$-independent variation of $\eta$), then from Eq. (3.9) we obtain

$$u(n) = -\frac{1}{4}\eta\left(1 + \frac{1}{16}\eta^2\right)\delta_{n,\pm 1} - \frac{1}{64}\eta^3\delta_{n,\pm 3} + \cdots \qquad (3.11)$$

which says, to lowest order in $\eta$, that the projected mean field state contains only those configurations in which the composite operators $\mathcal{C}^\dagger_{m,m+n}$ connect nearest neighbor sites. For an even length chain, there are three such configurations: two Majumdar-Ghosh-like alternating singlet states (these are absent for odd length chains), and the recently studied AKLT state![14,15] This observation adds support to the belief that the AKLT model has similar properties to the Heisenberg model which does not have the AKLT biquadratic term.

The mean field equations (3.5) and (3.6) lead to the following results for $c$ and $\kappa$:

*One dimension.*—At $T = 0$, we expand the solution to leading powers of $1/S$. Restoring the unit of energy which is given by the Heisenberg exchange coupling $J$, we obtain:

$$c = \left((S + 1 - \frac{1}{2}\right) \sqrt{2}J/\hbar[1 + O(1/S)] \quad (3.12)$$

$$\kappa = 16 \exp\left[-\pi\left(S + \frac{1}{2}\right)\right][1 + O(1/S)].$$

The antiferromagnetic correlations were shown to decay as $\exp(-\kappa R)$. Since true magnon excitations carry spin 1, they are bound states of pairs of Schwinger boson excitations. Thus we expect their gap to be $\Delta = c\kappa$, at the zone edge. This assignment is confirmed in the position of the peaks of the structure factor.[9] It is also a manifestation of the underlying Lorentz symmetry of the quantum antiferromagnet. The gap has the asymptotic form $\Delta \sim S \exp(-\pi S)$, which should be compared with the result $\Delta \sim S^2 \exp(-\pi S)$, obtained from the two loop order calculation of the $(1 + 1)$ dimensional sigma model.[4] It is remarkable that our simple mean field theory reproduces the asymptotic $S$-dependence of the Haldane gap. All is not well, however, because our mean field theory is unable to discern the topological terms responsible for the gaplessness of all half-odd-integer antiferromagnetic chains. Alternatively stated, the Lieb-Schultz-Mattis theorem,[6] which exploits the differing properties of integer and half-odd-integer spins under SU(2) rotations is violated at the mean field level, since it requires that all half-odd-integer Heisenberg antiferromagnetic chains must have either degenerate ground states or gapless excitations in the thermodynamic limit. We stress that the bosonic mean field theory *is* applicable to any model in which the ground state is ordered. In particular, this applies to the $S = 1/2$ model in two dimensions, as argued in the following discussion.

*Two dimensions.*—For $T < JS(S + 1)$ the solutions of Eqs. (3.5, 3.6) yield:

$$c = Z_c\sqrt{8JS}/\hbar \quad (3.13)$$

and,

$$\kappa = \exp[-Z_\kappa 2\pi S(S + 1)J/T]. \quad (3.14)$$

The solutions to the renormalization factors $Z_c(T, S)$, and $Z_\kappa(T, S)$ are obtained numerically for small values of $S$ in Table I. It was also previously shown[8] that Eq. (3.5) ensures that $Z_\kappa$ has a finite $T = 0$ limit for all $S > S_c \sim 0.2$, and is only weakly $T$ dependence for $T < JS(S + 1)$. For large $S$, $\lim_{S\to\infty} Z_\kappa = 1$, and Eq.

## TABLE I

| Theory | Coefficient | S = 1/2 | S = 1 |
|---|---|---|---|
| SB-MFT | $Z_c$ | 1.159 | 1.079 |
| SWT | $Z_c = 1 + 0.158/2S$ | 1.158 | 1.079 |
| SB-MFT | $Z_\chi$ | 0.53 ± 0.01 | 0.73 ± 0.01 |
| SWT | $Z_\chi = 1 - 0.552/2S$ | 0.448 | 0.724 |
| SB-MFT | $JS(S+1)(dZ_\chi/d_T)$ | 0.22 ± 0.01 | 0.27 ± 0.01 |
| SB-MFT | $Z_\kappa$ | 0.232 | 0.442 |
| RG-SWT | $Z_\kappa = \hbar c Z_c Z_\kappa / a\sqrt{8(S+1)}$ | 0.200 | 0.421 |
| SB-MFT | $\delta = C_v[T/S(S+1)J]^{-2}$ | 1.3 ± 0.05 | 1.2 ± 0.05 |

Resuls of the Schwinger boson mean field theory (SB-MFT) compared to spin wave theory[19] (SWT), and to the sigma model calculation[22] (CHN)⟩ $Z_c$, $Z_\chi$ and $Z_\kappa$, are the $T \to 0$ limit of the renormalization constants of the spin wave velocity, susceptibility and correlation length exponent respectively. (See Eqs. (3.13, 3.14)).

(3.12) agrees, to one loop order, with the renormalization group calculation of the classical Heisenberg model.[13,11] Since $Z_\kappa(S = 1/2) = 0.246$, it is apparent that quantum fluctuations drastically reduce the correlation length at finite temperatures from its classical value. On the other hand $\kappa^{-1}$ still diverges at $T = 0$, which implies that this system has a Neel ordered ground state, in agreement with numerical results for finite size systems.[16,17] Hirsch and Tang[18] have recently shown that by taking $T \to 0$ limit on a *finite* system, for $S > S_c$ Bose condensation occurs, since the $\mathbf{k} = (0, 0)$, $(\pi, \pi)$ modes get macroscopically occupied in order to satisfy Eq. (3.5). $S(\pi, \pi)$ was shown to agree with Anderson's spin wave calculation[2] of the ground state staggered magnetization squared. For $T > JS(S+1)$, a breakdown of the mean field theory occurs, and no solution for $\kappa$ and $c$ is found. This upper temperature does not correspond to a true phase transition (except perhaps for the large $N$ SU($N$) model), but to the breakdown of coherence between neighboring antiferromagnetically aligned spins.

Since, in effect, spin wave theory is a continuation of the SB-MFT to the broken symmetry phase, it is interesting to compare our $N = 2$ results to those obtained by the Hartree-Fock approximation of SWT.[19] 1) The values of the spin wave velocity renormalization $Z_c$, Eq. (3.13), agree well for $S = 1/2$ and $S = 1$, as shown in Table 1. Since $\kappa$ vanishes at $T = 0$, our quasiparticle dispersion matches the spin wave result. 2) The expression for $F^{MF} - E_c$ (where $E_c$ is the classical energy) is twice that of SWT. This is the same situation we have previously encountered in our results for the ferromagnetic chain in section 2. Here, however, we have not yet attempted to compute the gaussian corrections. 3) The spin correlation function $S^{MF}$ (3.7) is exactly 3/2 times the *rotationally averaged* expression of SWT. It can be easily verified that the susceptibility sum rule yields $\Sigma_{\mathbf{q},q_n} S^{MF} = S(S + 1)/2$ which is also 3/2 too large.

Therefore, in order to obey the condition that our theory *for* $N = 2$ should match SWT at $T = 0$, for large $S$, and also obey the sum rule, we correct our free energy and correlation functions by $F = \frac{1}{2} F^{MF}$, and $S(\mathbf{q}, \omega) \equiv \frac{2}{3} S^{MF}$. We suggest without proof that this normalization partly compensates for the fluctuation effecs, missed by the static mean field theory.

In Fig. 3, we plot the dynamical structure factor[9] in the positive $(\omega, \tilde{q})$ quadrant, where $\tilde{q} \equiv |\tilde{\mathbf{q}}|$. For $(\omega, \tilde{q}) < (T, T/c)$ here is a reflection symmet6ry on both energy and momentum axis. Two distinct regimes are observed: (a) $(\omega, \tilde{q}) \le (c\kappa, \kappa)$, and (b) $(\omega, \tilde{q}) \ll (c\kappa, \kappa)$. Region (a) is a quasielasic peak, which increases, and narrows with decreasing $\kappa$. This peak turns into the magnetic Bragg peak at $T = 0$, and its width reflects the overdamped nature of the spin-waves with wavelength longer than the coherence length. In region (b), the structure factor becomes asymptotically proportional to the naïve spin wave theory result, which predicts spin wave peaks at energy $c|\tilde{\mathbf{q}}|$. We note that there is a gap between the

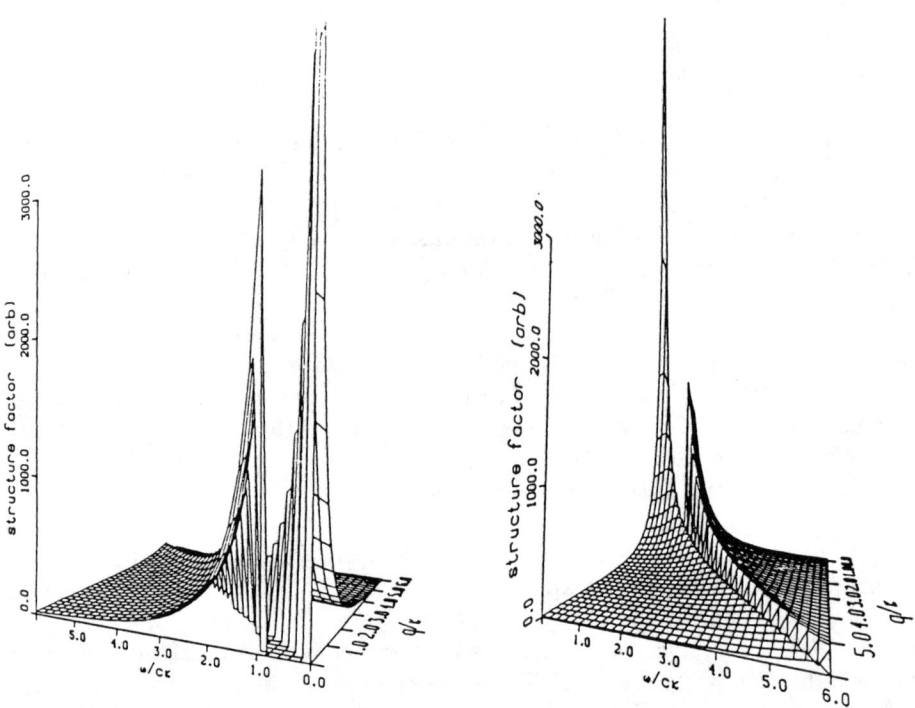

FIGURE 3 Two views of the structure factor of the square lattice antiferromagnet at $T < JS(S + 1)$. $\tilde{q}$ is the distance from the antiferromagnetic wave vector $(\pi, \pi)$. $\kappa$ is the inverse correlation length given in Eq. (3.14), and $c$ is the spin wave velocity Eq. (3.13). For low frequencies $\omega < T$, the structure factor is symmetric under reflection on both $\tilde{q}$ and $\omega$ axis.

normal ($\omega < c\bar{q}$) and the anomalous ($\omega > c\sqrt{\kappa^2 + q^2}$) contributions of the Schwinger boson scattering. We suspect, however, that this structure is an artifact of the static mean field approximation, and that it might be washed out by fluctuations in $\lambda$ and $Q$. Nevertheless, it would be interesting to see whether any double-peak features could be experimentally resolved.

It is easy to compute the uniform susceptibility,[9] which is given by

$$\chi = \frac{1}{T} S(\mathbf{q} = 0) = \frac{1}{3T} \int \frac{d^dk}{(2\pi)^d} n_k n_k + 1) \equiv \frac{1}{8J} Z_\chi \qquad (3.13)$$

In Fig. 4, we plot out $\chi(T)$ (for the applicable range of $T$) and show how it inerpolates between the rotationally averaged SWT result[19] and the high temperature series (HTS) expansion of Rushbrooke and Wood.[20] It is also important to note slight discrepancies for the value of $Z_\chi$ between our result and that of Oguchi as seen in Table I. The disagreement, probably arises from the order of

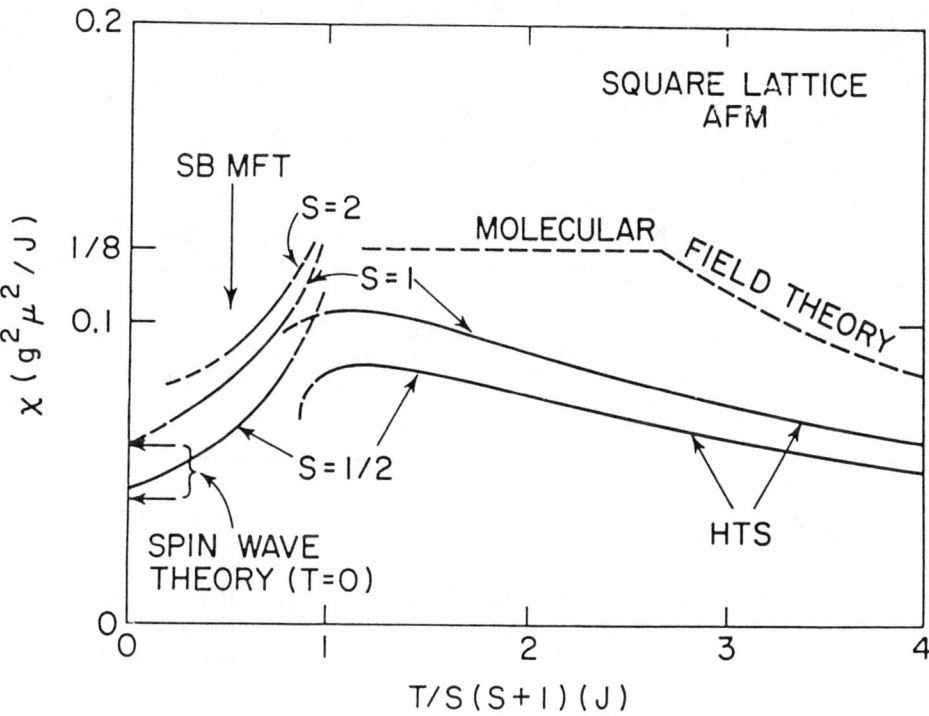

FIGURE 4 The uniform magnetic susceptibility for different spin value $S$. High temperature series (HTS) and the molecular field theory are given in Ref. [20]. The spin wave theory[19] susceptibility is rotationally averaged. The Schwinger boson mean field theory (SB-MFT) interpolates between two regimes.

limits in Eq. (3.13). If the zero temperature limit is taken first (say on a finite lattice), the first term of the integrand will be dropped. In the correct thermodyanmic limit, however, that integral contributes an additive constant to $\chi$, because of it's logarithmic dependence on the correlation length. Monte-Carlo calculations[21] of $\chi$ on finite lattices have found our theory in good agreement with the numerical results. We also present our result for the specific heat $T^2$ coefficient $\delta$ in Table I.

Chakravarty Halperin and Nelson[22] (CHN) have evaluated the temperature dependent correlation length $\kappa^{-1}$ of the 2-d Heisenberg model by studying the non-linear sigma model in a slab of finite thickness. They applied Oguchi's spin wave theory[19] (SWT) to determine the appropriate value of the sigma model coupling $g$ for the nearest neighbor model of $S = 1/2$. They found that $g < g_c$, and that the correlation length has renormalized classical behavior which agrees to the one loop order with our result Eq. (3.14). In table I, we compare CHN's renormalization constant to ours. The discrepancy in its numerical value could be traced back to their use of the $Z_\chi$ from spin wave theory. CHN have demonstrated the consistency of their $\kappa(T)$ with the experimentally determined correlation.[23] The regime of validity of this field theory, without the Berry phases measuring the hedgehog configurations, is identical to that of the SB-MFT. This was shown by Read and Sachdev[10] who generalized the mapping into the continuum theories for all SU($N$) antiferromagnetic models. These terms were shown to which lead to destructive interference between different topological sectors, and thus to quasidegeneracy of the disordered ground states.[7] Incorporating these effects in the SB-MFT is essential for the understanding of the ground state of the frustrated 2-d $S = 1/2$ antiferromagnet.

We thank the authors of Ref. [18,10,21] for sending us their manuscripts prior to publication. A. A. was supported by the Division of Material Sciences, U.S. Department of Energy, under contract No. DE-AC02-76H00016. D. P. A. was supported in part by grant No. DMR-MRL-85-19460.

## REFERENCES

1. N. D. Mermin and H. Wagner, *Phys. Rev. Lett.* **17**, 1144 (1966).
2. P. W. Anderson, *Phys. Rev.* **86**, 694 (1952).
3. H. Bethe, *Z. Physik* **71**, 205 (1931); L. Hulthén, *Ark. Mat. Ast. Fys.* **26A**, 1 (1938).
4. F. D. M. Haldane, *Phys. Rev. Lett.* **50**, 1153 (1983); *Phys. Lett. A* **93**, 464 (1983).
5. Recent generalizations of the LSM theorem have been discussed in Ian Affleck and Elliott H. Lieb, *Lett. Math. Phys.*, **12**, 57 (1986).
6. E. H. Lieb, T. Schultz, and D. C. Mattis, *Ann. Phys. (N.Y.)* **16**, 407 (1961).
7. F. D. M. Haldane, *Phys. Rev. Lett.* **61**, 1029 (1988).

8. D. P. Arovas and A. Auerbach, *Phys. Rev. B* **38**, 316 (1988).
9. A. Auerbach and D. P. Arovas, *Phys. Rev. Lett.* **61**, 617 (1988).
10. N. Read and Subir Sachdev, preprint, and these proceedings.
11. Minoru Takahashi, *Prog. Theor. Phys. Supp.* **87**, 233 (1986); *Phys. Rev. B* **36**, 2791 (1987).
12. In the original Takahashi theory, $Q$ is replaced by $S$ in the expression for $n_k$. At low temperature, $Q - S = \mathbb{O}(T^{3/1})$, and the quoted low-temperature expansions are identical. The inclusion of $Q$ rather than $S$ constitutes a refinement of Takahashi's basic theory as discussed in Ref. [11].
13. Stephen H. Shenker and Jan Tobochnik, *Phys. Rev. B* **22**, 4462 (1980).
14. Ian Affleck, Tom Kennedy, Elliott H. Lieb, and Hal Tasaki, *Phys. Rev. Lett.* **59**, 799 (1987).
15. Daniel P. Arovas, Assa Auerbach, and F. D. M. Haldane, *Phys. Rev. Lett.* **60**, 531 (1988).
16. J. Oitmaa and D. D. Betts, *Can. J. Phys.* **56**, 897 (1978).
17. J. D. Reger and A. P. Young, *Phys. Rev. B* **37**, 5978 (1988).
18. J. E. Hirsch and S. Tang, preprint.
19. T. Oguchi, *Phys. Rev.* **117**, 117 (1960).
20. G. S. Rushbrooke and P. J. Wood, *Mol. Phys.* **1**, 257 (1958).
21. Y. Okabe, M. Kikuchi and A. D. S. Nagi, preprint.
22. Sudip Chakravarty, B. I. Halperin, and D. Nelson, *Phys. Rev. Lett.* **60**, 1057 (1988).
23. G. Shirane, Y. Endoh, R. J. Birgeneau, M. A. Kastner, Y. Hidaka, M. Oda, M. Suzuki, and T. Muramaki, *Phys. Rev. Lett.* **59**, 1613 (1987).

P. N. Brusov
The Laboratory of the Theory of High Temperature Superconductivity
Rostov-on-Dov Physical Research Institute
USSR

and

Department of Physics and Astronomy
Northwestern University
Evanston, IL 60208-3112, U.S.A.

# The Theory of the Collective Excitations in the A-Phase of He$^3$

Within the model of superfluid He$^3$, created by a path integration method, the microscopic theory of collective excitations (CE) in He$^3$-A is constructed. The whole CE spectrum with taking into account the damping of CE is calculated. The cause of the additional Goldstone ($gd$) mode existence in the weak coupling approximation and their analogy to W-bosons in the weak interaction theory are discussed. The whole set of equations, which describe the CE in arbitrary magnetic fields is obtained. They are solved for small magnetic fields and the linear Zeeman effect for the clapping ($cl$) and pairbreaking ($pb$) modes is obtained.

The A-phase is maybe the most interesting object in superfluid He$^3$. It gives us an example of an anisotropic superfluid quantum liquid. The main features of He$^3$-A are connected with the existence on the Fermi surface of two nodes in the gap of a single particle spectrum. This leads to the existence in the system of chiral fermions, gauge fields, W- and Z-bosons, zero-charge phenomenon and to the damping of CE already at zero momentum of CE and to many other consequences.

In this paper we construct the microscopic theory of CE in He$^3$-A, using the path integration method.

## I. MODEL OF He$^3$

The model of He$^3$, which we describe below was first suggested by Alonso and Popov[1] and developed by Brusov and Popov.[2] We describe the He$^3$ system by the anticommuting functions $\chi_s(\vec{x}, \tau)$ and $\bar{\chi}_s(\vec{x}, \tau)$ with the Fourier expansion

$$\chi_s(\vec{x}, \tau) = \frac{1}{(\beta V)^{1/2}} \sum_{\vec{k},\omega} a_s(\vec{k}, \omega) \exp[i(\omega\tau - \vec{k}\vec{x})]. \tag{1}$$

Here $s = \pm$ is the spin index $\vec{x} \in V = L^3$, $\tau \in [0, \beta]$, $\beta = T^{-1}$ (in units $\hbar = k_B = 1$, $k_i = 2\pi n_i/L$, $\omega = (2n + 1)\pi/\beta$; $n$ and $n_i$ are integers). Let us consider the statistical sum for this system $\int \exp S \, d\bar{\chi} d\chi$, where the functional

$$S = \int_0^\beta d\tau \int d\vec{x} \sum_s \bar{\chi}_s(\vec{r}, \tau) \partial_\tau \chi_s(\vec{x}, \tau) - \int_0^\beta H'(\tau) d\tau \tag{2}$$

has the meaning of the action corresponding to the Hamiltonian

$$H'(\tau) = \int d\vec{x} \sum_s \left( \frac{1}{2m} \nabla \bar{\chi}_s(\vec{x}, \tau) \nabla \chi_s(\vec{x}, \tau) - (s\mu_0 H + \lambda) \bar{\chi}_s(\vec{x}, \tau) \chi_s(\vec{x}, \tau) \right.$$

$$\left. + \frac{1}{2} \int dxdy u(\vec{x} - \vec{y}) \sum_{s,s'} \bar{\chi}_s(\vec{x}, \tau) \bar{\chi}_{s'}(\vec{y}, \tau) \chi_{s'}(\vec{y}, \tau) \chi_s(\vec{x}, \tau), \right. \tag{3}$$

in which $\lambda$ is the chemical potential, $\mu_0$ is the magnetic moment of the Fermi particle, and $H$ is the magnetic field.

Using the idea that only fermions in the vicinity of the Fermi surface are important for the superfluidity we divide the Fermi-fields (1) into two parts: "fast" fields and "slow" ones. "Fast" fields $\chi_1$ and $\bar{\chi}_1$ are determined by the part of expansion (1) with $|k - k_F| > k_0$ or $|\omega| > \omega_0$, and "slow" ones $\chi_0$ and $\bar{\chi}_0$ are equal $\chi_0 = \chi - \chi_1$, $\bar{\chi}_0 = \bar{\chi} - \bar{\chi}_1$.

The auxiliary parameters $k_0$ and $\omega_0$ are defined only accurate to their order of magnitude, and the physical results should not depend on their concrete choice.

Than we integrate first over the "fast" Fermi fields and then over the "slow" ones, using during these two stages different perturbation-theory schemes. The integral over the fast fields $\chi_1$ and $\bar{\chi}_1$ will be written in the form

$$\int \exp S \, d\bar{\chi}_1 d\chi_1 = \exp \tilde{S}[\bar{\chi}_0, \chi_0]. \tag{4}$$

The functional $\tilde{S}$ has the meaning of the action of the "slow" fields $\chi_0$ and $\bar{\chi}_0$ for which $|k - k_F| < k_0$ and $|\omega| < \omega_0$. The general form of the functional $\tilde{S}$ is a sum of functionals of the even powers in the fields $\chi_0$ and $\bar{\chi}_0$:

$$\tilde{S} = \sum_{n=0} \tilde{S}_{2_n} \tag{5}$$

Neglecting the higher functionals $\tilde{S}_6$, $\tilde{S}_8$, ... and omitting the constant $\tilde{S}_0$, which is no longer significant, we examine the forms of $\tilde{S}_2$ and $\tilde{S}_4$. The form of $\tilde{S}_2$

# The Theory of the Collective Excitations in the A-Phase of He³

corresponds to non-interacting quasiparticles near the Fermi surface, and is given by

$$\sum_{\vec{k},\omega,s} \epsilon_s(\vec{k},\omega) a_s^+(\vec{k},\omega) a_s(\vec{k},\omega), \quad |k - k_F| < k_0, \quad |\omega| < \omega_0, \tag{6}$$

with

$$\epsilon_s(\vec{k},\omega) \approx Z^{-1}(i\omega - c_F(k - k_F) + s\mu H). \tag{7}$$

Here, assuming that $\epsilon_s(\omega = 0, k = k_F, H = 0) = 0$, we have expanded $\epsilon_s$ in powers $\omega$, $k - k_F$ and $H$ and retained only the linear terms. The coefficient $c_F$ has the meaning of the velocity on the Fermi surface, $\mu$ is the magnetic moment of the quasiparticle, and $Z$ is a normalization constant.

The form $\tilde{S}_4$ describes the interaction of the quasiparticles and is given by

$$-\frac{1}{\beta V} \sum_{p_1+p_2=p_3+p_4} t_0(p_1, p_2, p_3, p_4) a_+^+(p_1) a_-^+(p_2) a_-(p_4) a_+(p_3)$$

$$-\frac{1}{2\beta V} \sum_{p_1+p_2=p_3+p_4} t_1(p_1, p_2, p_3, p_4)[2a_+^+(p_1) a_-^+(p_2) a_-(p_4) a_+(p_3)$$

$$+ a_+^+(p_1) a_+^+(p_2) a_+(p_4) a_+(p_3) + a_-^+(p_1) a_-^+(p_2) a_-(p_4) a_-(p_3)]. \tag{8}$$

Here $p = (\vec{k}, w)$, is the 4-momentum; $t_0(p_i)$ and $t_1(p_i)$ are respectively the symmetrical and antisymmetrical scattering amplitudes under the permutations $p_1 \leftrightarrows p_2$ and $p_3 \leftrightarrows p_4$. In the vicinity of the Fermi sphere we can put $\omega_i = 0$, $\vec{k}_i = k\vec{n}_i$ ($i = 1, 2, 3, 4$), where $\vec{r}_i$ are unit vectors such that $\vec{n}_1 + \vec{n}_2 = \vec{n}_3 + \vec{n}_4$. The amplitudes $t_0$ and $t_1$ should depend only on two invariants, for example on $(\vec{n}_1, \vec{n}_2)$ and $(\vec{n}_1 - \vec{n}_2, \vec{n}_3 - \vec{n}_4)$ with $t_0$ even and $t_1$ odd in the second invariant. We therefore have the expressions

$$t_0 = f((\vec{n}_1, \vec{n}_2); (\vec{n}_1 - \vec{n}_2, \vec{n}_3 - \vec{n}_4)),$$

$$t_1 = (\vec{n}_1 - \vec{n}_2, \vec{n}_3 - \vec{n}_4) g((\vec{n}_1, \vec{n}_2), (\vec{n}_1 - \vec{n}_2, \vec{n}_3 - \vec{n}_4)). \tag{9}$$

Here $t_0$ and $t_1$ are expressed in terms of the functions $f$ and $g$, which are even in the second argument.

The functional $\tilde{S}_2 + \tilde{S}_4$, defined by formulas (6)–(9) is the most general expression describing Fermi quasiparticles and their pair interaction near the Fermi sphere. The method of obtaining this functional in the path-integral formalism, and its investigation that follows below, constitute an alternative approach to that developed in the Landau theory of the Fermi liquid.

The functions $f$ and $g$ can be easily calculated for the gas model. For high-density systems they must be determined from experiment.

We consider hereafter a model with

$$f = 0, \quad g = \text{const} < 0 \tag{10}$$

as the simplified model of He$^3$ with pairing in the $p$ state.

Using the Fermi-fields (1) for description of He$^3$ is the most logical, however significant difficulties appear when we use them to describe the low energy (infrared) phenomena in superfluid He$^3$. The cause of this is the absence of the singularities in the single particle Green function $\langle 0|T(\chi(\tilde{x}, \tau)\bar{\chi}(\tilde{y}, \tau_1))|0\rangle$ below $E = \Delta$, where $\Delta$ is a gap in a single particle spectrum $E(\vec{k}) = (\xi^2(\vec{k}) + \Delta^2)^{1/2}$, $\xi(\vec{k}) = c_F(k - k_F)$. Thus the description of infrared phenomena (with $E \ll \Delta$) such as zero sound, spin waves and so on is complicated in terms of Fermi-fields. We need to sum up an infinite set of the Feynman diagrams for a simple understanding of these phenomena. But there are Green functions which describe such kind of excitations directly, they are

$$\langle 0|T[\bar{\chi}(\tilde{x}, \tau)\chi(\tilde{x}, \tau); \bar{\chi}(\tilde{y}, \tau_1)\chi(\tilde{y}, \tau_1)]|0\rangle$$

$$\left\langle 0 \left| T \left[ \bar{\chi}(\tilde{x}, \tau) \frac{\sigma_a}{2} \chi(\tilde{x}, \tau); \bar{\chi}(\tilde{y}, \tau_1) \frac{\sigma_a}{2} \chi(\tilde{y}, \tau_1) \right] \right| 0 \right\rangle \tag{11}$$

Singularities, which appear in such complicated Green functions and which are absent in single particle ones are called CE. The most logical way for their description is the passage from Fermi-fields to Bose-fields which describe the Cooper pairs of quasifermions.

To realize such passage we introduce under the sign of the integral over the Fermi field a Gaussian integral of $\exp(\bar{c}\hat{A}c)$ with respect to the Bose field $c$, where $\bar{c}\hat{A}c$ is a quadratic form with a certain operator $\hat{A}$. We then shift the Bose field by a quadratic form of the Fermi fields, so as to annihilate the form $\tilde{S}_4$, of fourth-degree in the Fermi fields. The integral over the Fermi fields is then transformed into a Gaussian integral and is equal to the determinant of the operator $\hat{M}(c, \bar{c})$ that depends on the Bose fields $c$ and $\bar{c}$. We arrive at the functional

$$S_h = \bar{c}\hat{A}c + \ln \det/[\hat{M}(c, \bar{c})\hat{M}(0, 0)], \tag{12}$$

in which the ln det has been regularized by dividing $\hat{M}(c, \bar{c})$ by the operator $\hat{M}(0, 0) = \hat{M}(c, \bar{c})|_{c=\bar{c}=0}$.

The functional $S_h$ is called the "hydrodynamic action functional." It defines the point of the phase transition of the initial Fermi system as Bose condensation of the fields $c$ and $\bar{c}$, and determines the density of the condensate at $T < T_c$ and the spectrum of the collective excitations.

# The Theory of the Collective Excitations in the A-Phase of He³

In case of $p-$ pairing we need to introduce under the sign of the integral over the Fermi fields a Gaussian integral over the complex functions $c_{ia}(x, \tau)$ and $\bar{c}_{ia}(x, \tau)$ with the vector index $i$ and the isotopic index $a$ ($i, a = 1, 2, 3$). The Gaussian integral is of the form

$$\int d\bar{c}_{ia} dc_{ia} \exp\left(\frac{1}{g} \sum_{p,i,a} \bar{c}_{ia} + (p)c_{ia}(p)\right). \tag{13}$$

where $g$ is the constant (10). It is easily verified that the shift

$$c_{i1}(p) \to c_{i1}(p) + \frac{g}{2(\beta V)^{1/2}} \sum_{p_1+p_2=p} (n_{1i} - n_{2i})[a_+(p_2)a_+(p_1) - a_-(p_2)a_-(p_1)],$$

$$c_{i2}(p) \to c_{i2}(p) + \frac{gi}{2(\beta V)^{1/2}} \sum_{p_1+p_2=p} (n_{1i} - n_{2i})[a_+(p_2)a_+(p) + a_-(p_2)a_-(p_1)],$$

$$c_{i3}(p) \to c_{i3}(p) + \frac{g}{(\beta V)^{1/2}} \sum_{p_1+p_2=p} (n_{1i} - n_{2i})a_-(p_2)a_+(p_1) \tag{14}$$

does indeed eliminate the form $\tilde{S}_4$.

To calculate the Gaussian integral over the Fermi fields, we introduce a column $\psi_a(p)$ with elements

$$\psi_1(p) = a_+(p), \quad \psi_2(p) = -a_-(p), \quad \psi_3(p) = a_-^+, \quad \psi_4(p) = a_+^+(p) \tag{15}$$

and write down a quadratic form in the Fermi fields in the form

$$K = \frac{1}{2} \sum_{p_1+p_2,a,b} \psi_a^+(p) M_{a,b}(p_1, p_2) \psi_b(p_2). \tag{16}$$

The fourth-order matrix $M(p_1, p_2)$ with elements $M_{ab}(p_1, p_2)$ is given by

$$M = \begin{pmatrix} Z^{-1}(i\omega - \xi + \mu(\vec{H}\vec{\sigma}))\delta_{p_1,p_2} & \frac{1}{(\beta V)^{1/2}}(n_{1i} - n_{2i})c_{ia}(p_1+p_2)\sigma_a \\ -\frac{1}{(\beta V)^{1/2}}(n_{1i} - n_{2i})c_{ia}^+(p_1+p_2)\sigma_a & Z^{-1}(i\omega + \xi + \mu(\vec{H}\vec{\sigma}))\delta_{p_1 p_2} \end{pmatrix}, \tag{17}$$

where $\sigma_a$ ($a = 1, 2, 3$) are $2 \times 2$ Pauli matrices.

Integrating over the Fermi fields

$$\int e^K d\bar{\chi}_0 d\chi_0 = (\det \hat{M})^{1,2}, \tag{18}$$

we arrive at the "hydrodynamic action" functional in the form

$$S_h = \frac{1}{g}\sum_{p,i,a} c_{ia}^+(p)c_{ia}(p) + \frac{1}{2}\ln\det\frac{\hat{M}(c,\bar{c})}{\hat{M}(0,0)}. \tag{19}$$

This functional contains all the information on the physical properties of the system. In particularly it determines the transition temperature into the superfluid state, the order parameters ($OP$) of the superfluid states, the gap equation, the $CE$ spectrum and much else. We will study the superfluid states, the gap equation, the $CE$ spectrum and much else. We will use this to investigate the $CE$ spectrum.

## II. THE CE SPECTRUM IN THE ABSENCE OF MAGNETIC FIELDS

The collective modes ($CM$) in He$^3$-A describe the oscillations $\delta A_{i\alpha}$ of the $OP$ (complex $3\times 3$ matrix) around its equilibrium state

$$A_{i\alpha}^{(o)} = \Delta_o(e_1^i + ie_2^i)d_\alpha, \tag{20}$$

The unit vectors $\vec{e}_1$ and $\vec{e}_2$ describe the orbital part of the order parameter; their vector product determines the orbital anisotropy vector $\vec{l} = [\vec{e}_1, \vec{e}_2]$; and the unit vector $\vec{d}$ specifies the spin axis, i.e., the axis of the magnetic anisotropy.

The number of $CM$ in He$^3$-A as well as in any other phase equals 18 ($3\times 3\times 2$). In general all of these modes could be observed either in $NMR$ or in zero sound experiments.

The classification of $CM$ in He$^3$-A has been done by Volovik and Khazan[3] in terms of the irreducible representations of the symmetry group $H$ of the equilibrium state (20). In distinction from the modes in $^3$He-B which are characterized by one quantum number $J$ and a single parity (with respect to complex conjugation), the modes in He$^3$-A are characterized by two quantum numbers: $Q$ and $S_z$, and two parities $P^1$ and $P^2$. The charges $Q$ and $S_z$ assume the values 0, $\pm 1$, $\pm 2$, and 0, $\pm 1$, respectively. Owing to the parities $P^1$ and $P^2$ those modes which differ in the sign of either $S_z$ or $Q$ turn out to be degenerate. Consequently, if the wave vector $\vec{k}$ is parallel to the orbital anisotropy axis the spectrum of modes will consist of two four-fold degenerate branches.

An additional degeneracy of the spectrum of modes is exhibited in the weak coupling limit, on account of an enlargement of the group $H$ owing to hidden symmetry. This leads in particular to the appearance of four additional $gd$-modes, which were first obtained by Brusov and Popov.[4]

The calculation of the $CE$ spectrum was made in numerous papers (see, for example Ref. 5), where the energies of cl $E = 1.22\Delta_o$, flapping $(fl)E = 1.56\Delta_o$,

# The Theory of the Collective Excitations in the A-Phase of He³

and $pb$ $E = 2\Delta_o$ modes were obtained (here $\Delta_o$ is the maximum value of the gap $\Delta = \Delta_o \sin\theta$). These energy values were obtained without taking any $CE$ damping into account. But it's clear that vanishing of gap along the orbital anisotrophy axis $\vec{l}$ leads to the damping of $CE$ due to the decay into two fermions, because $CE$ with nonzero energy and small momentum can always decay kinematically into two fermions whose momenta are almost opposite and close to the axis $\vec{l}$. The whole $CE$ spectrum, taking into account this damping, was obtained first by Brusov and Popov.[2]

Following their paper we describe below the whole $CE$ spectrum in He³-A without magnetic fields. For calculating the $CE$ spectrum in the region $T_c - T \sim T_c$, we expand the functional $\ln \det$ in (19) in powers of the deviation $c_{ia}(p)$ from the condensate value $c_{ia}^{(0)}(p)$, which is different for different phases. We apply the shift $c_{ia}(p) \to c_{ia}^{(0)}(p) + c_{ia}(p)$ and separate from $S_h$ the quadratic form

$$\sum_p c_{ia}^+(p)c_{jb}(p)A_{ijab}(p) + \frac{1}{2}\sum_p (c_{ia}(p)c_{jb}(-p) + c_{ia}^+(-p)c_{jb}^+(-p))B_{ijab}(p). \quad (20)$$

This form determines, to first approximation, the Bose spectrum obtained from the equation

$$\det Q = 0. \quad (21)$$

Here $Q$ is a matrix of quadratic form, determined by the coefficient tensors $A_{ijab}$, $B_{ijab}$ in (20). These quantities are proportional to the integrals of the products of the Green's functions of the fermions. Most effective in the calculation of these integrals is the Feynman procedure customarily used in relativistic quantum theory. In the present case the procedure is based on the identity

$$\frac{1}{(\omega_1^2 + \xi_1^2 + \Delta^2)(\omega_2^2 + \xi_2^2 + \Delta^2)}$$

$$= \int \frac{d\alpha}{[\alpha(\omega_1^2 + \xi_1^2 + \Delta^2) + (1-\alpha)(\omega_2^2 + \xi_2^2 + \Delta^2)]^2} \quad (22)$$

It is easy to evaluate by this procedure the integrals with respect to the variables $\omega$ and $\xi$, and then with respect to the angle variables and the parameter $\alpha$. The quadratic part of $S_h$ for the $A$ phase of the model is a sum of three quadratic forms, the first of which depends on the variables $c_{i1}$, the second on $c_{i2}$, and the third on $c_{i3}$. The second and third form are transformed into the first by the substitutions $c_{i2} \to c_{i1}$ and $c_{i3} \to ic_{i2}$. The quadratic form of the variables $c_{i1}$ is

$$\sum_p \left\{ c_{i1}^+(p)c_{j1}(p)\left[\frac{\delta_{ij}}{g} + \frac{4Z^2}{\beta V}\sum_{p_1+p_2=p} n_{1i}n_{1j}G_1G_2(\xi_1 + i\omega_1)\cdot(\xi_2 \right.\right.$$

$$+ i\omega_2)\Big] + (c_{i1}^+(p)c_{j1}^+(-p) + c_{i1}(p)c_{j1}(-p)) \frac{2\Delta_0^2 Z^2}{\beta V} \sum_{p_2+p_2=p} (n_1 \pm in_2)^2 n_{1i} n_{1j} G_1 G_2 \Big\}, \tag{23}$$

where

$$G(p) = (\omega^2 + \xi^2 + \Delta^2)^{-1}, \qquad \Delta^2 = \Delta_0^2(n_1^2 + n_2^2) = \Delta_0^2 \sin^2\theta. \tag{24}$$

Here $\Delta_0$ is the maximum value of the energy gap of the Fermi spectrum. In the term $(n_1 \pm in_2)^2$ of (23) the upper and lower signs are taken when the multiplication is by $c_{i1}^+ c_{j1}^+$ and by $c_{i1} c_{j1}$, respectively.

We investigate now all the Bose-spectrum branches defined by (23) at zero momentum $\vec{k}$. At $\vec{k} = 0$ the form of the variables $c_{i1}(\omega, \vec{k} = 0)$, and $c_{i1}^+(\omega, \vec{k} = 0)$ is a sum of a form of $c_{11}$, $c_{11}^+$, $c_{21}$, $c_{21}^+$ and a form of $c_{31}$, $c_{31}^+$. The functions that are the coefficients of $c_{i1}^+ c_{j1}$, $c_{i1} c_{j1}^+$, and $c_{i1} c_{j1}$ $(i, j = 1, 2)$ can be expressed in the form

$$\frac{\delta_{ij}}{g} + \frac{4Z^2}{\beta V} \sum_{p_1+p_2=p} n_{1i} n_{1j} (\xi_1 + i\omega_1)(\xi_2 + i\omega_2) G_1 G_2$$

$$= \frac{2\delta_{ij} Z^2}{\beta V} \sum_{p_1+p_2=p} (n_1^2 + n_2^2)[(\xi_1 + i\omega_1)(\xi_2 + i\omega_2)G_1 G_2 - G_1], \tag{25}$$

$$= \frac{2\Delta_0^2 Z^2}{\beta V} \sum_{p_2+p_2=p} (n_1 \pm in_2)^2 n_{1i} n_{1j} G_1 G_2 = b_{ij} \frac{Z^2 \Delta_0^2}{2\beta V} \sum_{p_1+p_2=p} (n_1^2 + n_2^2) G_1 G_2$$

Here $b_{ij}$ $(i, j = 1, 2)$ are the elements of the matrix

$$\begin{pmatrix} 1 & \pm i \\ \pm i & -1 \end{pmatrix} \tag{26}$$

in which the minus sign corresponds to the variables $c_{1i} c_{1j}$, and the plus sign to $c_{1i}^+ c_{1j}^+$.

On going from the left to the right sides of the formulas in (25) we used the possibility of averaging (at $\vec{k} = 0$) over the azimuthal angle, of which the functions $G_1$, $G_2$, $\xi_1$ and $\xi_2$ are independent. We have used also the inequality

$$\frac{\delta_{ij}}{g} + \frac{Z^2}{\beta V} \sum_{p_1} n_{1i} n_{1j} G_1 = 0 \tag{27}$$

which determines the value of the gap that enters in $G_1 = (\omega_1^2 + \xi_1^2 + \Delta_1^2 \sin^2 \theta_1)^{-1}$.

We denote the coefficient of $\delta_{ij}$ in (25) by $f(\omega)$, and the coefficient of $b_{ij}$ by $g(\omega)$. We put also

$$u_1 = \operatorname{Re} c_{11}, \quad v_1 = \operatorname{Im} c_{11}, \quad u_2 = \operatorname{Re} c_{21}, \quad v_2 = \operatorname{Im} c_{21}. \tag{28}$$

The quadratic form of the variables $u_1, u_2, v_1, v_2$ ($\vec{k} = 0$) can then be expressed as a sum of two forms:

$$[(f(\omega) + g(\omega))(u_1^2 + v_2^2) - 2g(\omega)u_1 v_2] + [(f(\omega)$$
$$- g(\omega)(v_1^2 + u_2^2) - 2g(\omega)v_1 u_2]. \tag{29}$$

These forms correspond to the matrices

$$\begin{pmatrix} f(\omega) + g(\omega) & -g(\omega) \\ -g(\omega) & f(\omega) + g(\omega) \end{pmatrix}, \begin{pmatrix} f(\omega) - g(\omega) & -g(\omega) \\ -g(\omega) & f(\omega) - g(\omega) \end{pmatrix} \tag{30}$$

Equating to zero the determinants of the matrices (30), we obtain the equations

$$f(\omega)(f(\omega) + 2g(\omega)) = 0, \qquad f(\omega)(f(\omega) - 2g(\omega)) = 0$$

or (31)

$$f(\omega) = 0, \quad f(\omega) + 2g(\omega) = 0, \quad f(\omega) - 2g(\omega) = 0.$$

We add to (31) the equation obtained from an examination of the terms with $c_{31}$ and $c_{31}^+$:

$$h(\omega) = g^{-1} + \frac{4Z^2}{\beta V} \sum_{p_1+p_2=p} n_3^2(\xi_1 + i\omega_1)(\xi_2 + i\omega_1)G_1 G_2$$

$$= \frac{2Z^2}{\beta V} \sum_{p_1+p_2=p} \{2n_3^2(\xi_1 + i\omega_1)(\xi_2 + i\omega_1)G_1 G_2 - (n_1^2 + n_2^2)G_1\} = 0. \tag{32}$$

The three equations of (31) can be combined into one:

$$\frac{2Z^2}{\beta V} \sum_{p_1+p_2=p} (n_1^2 + n_2^2)[((\xi_1 + i\omega_1)(\xi_2 + i\omega_2) \pm (1, 0)\Delta^2)G_1 G_2 - G_1] = 0, \tag{33}$$

in which $\pm(1, 0)\Delta^2$ denotes either $\Delta^2$ or $-\Delta^2$ or 0.

Changing over (at $T = 0$) in (32) and (33) from the sums to (32) and (33) in the form

$$\frac{2Z^2k_F^2}{(2\pi)^4c_F}\int d\Omega d\omega_1 d\xi_1 \left[\frac{2\cos^2\theta(\xi_1+i\omega_1)(\xi_2+i\omega_2)}{(\omega_1^2+\xi_1^2+\Delta^2)(\omega_2^2+\xi_2^2+\Delta^2)} - \frac{\sin^2\theta}{(\omega_1^2+\xi_1^2+\Delta^2)}\right] = 0$$

$$\frac{2Z^2k_F^2}{(2\pi)^4c_F}\int \sin^2\theta d\Omega d\omega_1 d\xi_1 \left[\frac{(\xi_1+i\omega_1)(\xi_2+i\omega_2)\pm(1,0)\Delta^2}{(\omega_1^2+\xi_1^2+\Delta^2)(\omega_2^2+\xi_2^2+\Delta^2)}\right.$$

$$\left. - \frac{1}{(\omega_1^2+\xi_1^2+\Delta^2)}\right] = 0 \quad (34)$$

Taking the integrals with respect to $\omega_1$ and $\xi_1$ with the aid of Feynman techniques, we get

$$\frac{Z^2k_F^2}{4\pi^3c^F}\int_0^1 d\alpha \int \cos\theta\, d\Omega\left[\ln\frac{\Delta^2}{\Delta^2+\alpha(1-\alpha)\omega^2} - \frac{\alpha(1-\alpha)\omega^2}{\Delta^2+\alpha(1-\alpha)\omega^2}\right] = 0$$

$$\frac{Z^2k_F^2}{4\pi^3c_F}\int_0^1 d\alpha \int \sin^2\theta\, d\Omega\left[\ln\frac{\Delta^2}{\Delta^2+\alpha(1-\alpha)\omega^2}\right. \quad (35)$$

$$\left. - \frac{2\alpha(1-\alpha)\omega^2+\Delta^2\mp(1,0)\Delta^2}{\Delta^2+\alpha(1-\alpha)\omega^2}\right] = 0.$$

Calculating the integrals with respect to $\alpha$, substituting $\omega \to \Delta_0\omega$, and putting $\cos\theta = x$, we arrive at the equations

$$\int_0^1 dx(1-x^2)\frac{\omega^2+4(1-x^2)}{\omega[\omega^2+4(1-x^2)]^{1/2}}\ln\frac{[\omega^2+4(1-x^2)]^{1/2}+\omega}{[\omega^2+4(1-x^2)]^{1/2}-\omega} = 0,$$

$$\int_0^1 dx(1-x^2)\frac{\omega^2+2(1-x^2)}{\omega[\omega^2+4(1-x^2)]^{1/2}}\ln\frac{[\omega^2+4(1-x)]^{1/2}+\omega}{[\omega^2+4(1-x^2)]^{1/2}-\omega} = 0,$$

$$\int_0^1 dx(1-x^2)\frac{\omega}{[\omega^2+4(1-x^2)]^{1/2}}\ln\frac{[\omega^2+4(1-x^2)]^{1/2}+\omega}{[\omega^2+4(1-x^2)]^{1/2}-\omega} = 0, \quad (36)$$

$$\int_0^1 dxx\frac{\omega^2+2(1-x^2)}{\omega[\omega^2+4(1-x^2)]^{1/2}}\ln\frac{[\omega^2+4(1-x^2)]^{1/2}+\omega}{[\omega^2+4(1-x^2)]^{1/2}-\omega} = 0,$$

The first of these equations is the equation $f - 2g = 0$, the second is $f = 0$, and the third is $f + 2g = 0$, and the fourth is $h = 0$. They determine the Bose spectrum at $\vec{k} = 0$ following the analytic continuation $i\omega \to E$. The spectrum branches corresponding to the second and fourth equations are doubly degenerate. To take into account the forms of the variables $c_{i2}$ and $c_{i3}$ which lead to similar equations for the spectrum, it is necessary to multiply by 3 the multiplicity of each branch in the considered model.

The third and fourth equations in (36) have roots $\omega = 0$ and correspond to the $gd$-modes. From the first and second equation we obtain the complex energies of the nonphonon modes $E_1(\vec{k} = 0)$ and $E_2(\vec{k} = 0)$.

The result takes the form

$$E_1(0) = \Delta_0(1.96 - i0.31), \quad E_2(0) = \Delta_0(1.17 - i0.13), \tag{37}$$

the second of the modes being doubly degenerate.

The difference between Re $E$ here[4] and in Ref. 5 is connected with the fact that taking $CE$ damping into account (Im $E \neq 0$) leads via dispersion relations to renormalization of Re $E$. The other interesting fact, obtained first in Ref. 4, is that the number of $gd$ modes in the weak coupling approximation is equal to 9 rather than 5, which takes place in real He$^3$-A. The existence of 4 additional quasi- $gd$ spin-orbit modes is the consequences of the *latent* symmetry of the system and we investigate below this equation in more detail.

## III. THE LATENT SYMMETRY, ADDITIONAL $gd$-MODES, W-BOSONS

We shall show that taking into account the strong coupling effects decreases the number of phonon modes from 9 to 5, and that turning-on a magnetic field decreases the number of phonon modes from 9 to 6 for weak coupling and from 5 to 4 when strong coupling effects are taken into account.

We consider in the Ginzburg-Landau region $|T - T_c| \ll T_c$, that part $F$ of the action which is independent of the gradients. In the weak coupling model we have

$$F = -\operatorname{tr} AA^+ + v\operatorname{tr} A^+AP + (\operatorname{tr} AA^+)^2 + \operatorname{tr} AA^+AA^+$$

$$+ \operatorname{tr} AA^+A^*A^T - \operatorname{tr} AA^TA^*A^+ - 1/2 \operatorname{tr} AA^T \operatorname{tr} A^+A^*, \tag{38}$$

where $A$ (the order parameter) is a complex matrix with elements $A_{ia}$. The A-phase in the weak coupling is described by the order parameter

$$\frac{1}{2}\begin{pmatrix} 1 & 0 & 0 \\ i & 0 & 0 \\ 0 & 0 & 0 \end{pmatrix}, \tag{39}$$

and the phonon variables are

$$u_{21} - v_{11}, u_{12} + v_{22}, \quad u_{13} + v_{23}, u_{31}, v_{31}, u_{32}, v_{32}, u_{33}, v_{33}, \tag{40}$$

where $u_{ia} = \text{Re } A_{ia}$ and $v_{ia} = \text{Im } A_{ia}$. These variables correspond to the phonon branches of the spectrum not only in the Ginzburg-Landau region, but also at all $T < T_c$. In the limit as $T \to 0$, the first three of the variables in (40) correspond to sound waves with $c_F k/\sqrt{3}$, and the six remaining to orbital waves $c_F k_\parallel$. The phonon spectrum is thus degenerate in the spin index.

To take into account the strong coupling effects, we consider $F$ with arbitrary coefficients of the fourth-order terms:

$$F = -\text{tr } A^+A + \nu \text{ tr } A^+AP + a(\text{tr } A^+A)^2 + b \text{ tr } AA^+AA^+$$
$$+ c \text{ tr } AA^+A^*A^T + d \text{ tr } AA^TA^*A^+ + e \text{ tr } AA^T \text{ tr } A^+A^*. \quad (41)$$

The condition $\delta F = 0$ yields in the $A$-phase an order parameter in the form

$$\frac{1}{2}(a + b + d)^{1/2} \begin{matrix} 1 & 0 & 0 \\ i & 0 & 0 \\ 0 & 0 & 0 \end{matrix}. \quad (42)$$

To find the phonon variables, we calculate the second variation $\delta^2$

$$\delta^2 F = -\text{tr } AA^+ + \nu \text{ tr } A^+AP + a \text{ tr}[(A^+C)^2 + (C^+A)^2 + 2A^+AC^+C$$
$$+ 2A^+CC^+A] + b \text{ tr}[2AA^+CC^+ + 2A^+AC^+ + C + AC^+AC^+$$
$$+ A^+CA^+C] + c \text{ tr}[AA^+CC^T + A^+A^*C^TC + A^*A^TCC^+ \quad (43)$$
$$+ A^TAC^+C^* + AC^+A^*C^T + A^TCA^+C^*] + d \text{ tr}[AA^TC^*C^+$$
$$+ A^TA^*C^+C + A^*A^+CC^+ + A^+AC^TC^*] + 4e|\text{tr } C^TA|^2$$

where $C$ is the matrix (42) and $A$ is a variable matrix. Substituting the values of $C$, $C^+$, $C^*$, and $C^T$, we get

$$\delta^2 F = \nu(a + b + d)(u_{13}^2 + v_{13}^2 + u_{23}^2 + v_{23}^2 + u_{33}^2 + v_{33}^2)$$
$$+ 4a(u_{11} + v_{21})^2 + 2b[2(u_{11} + v_{21})^2 - (u_{13} - v_{23})^2$$
$$- (u_{12} - v_{22})^2 - (u_{22} + v_{12})^2 - (u_{23} + v_{13})^2$$
$$- 2(u_{32}^2 + v_{32}^2 + u_{33}^2 + v_{33}^2)]$$
$$+ 2c[2(u_{11} - v_{21})^2 + (u_{23} + v_{13})^2 + 2(u_{21} + v_{11})^2$$
$$+ (u_{21} - v_{22})^2 + (u_{22} + v_{12})^2 + (u_{13} - v_{23})^2] \quad (44)$$

$$+ 2d[2(u_{11} + v_{21})^2 - (u_{12} - v_{22})^2 - (u_{22} + v_{12})^2$$

$$- (u_{13} - v_{23})^2 - 2(u_{22} - v_{12})^2$$

$$- 2(u_{23} - v_{13})^2 - (u_{23} + v_{13})^2 - 2(u_{32}^2 + v_{32}^2) + u_{33}^2 + v_{33}^2)]$$

$$+ 4e[(u_{11} - v_{21})^2 + (u_{21} + v_{11})^2].$$

We consider first the system in a zero magnetic field ($\nu = 0$). The (44) is the sum of five quadratic forms multiplied by the independent coefficients $a$, $b$, $c$, $d$, and $e$. The variables

$$u_{12} + v_{22}, u_{13} - v_{23}, u_{21} - v_{11}, u_{31}, v_{31} \tag{45}$$

do not enter in any of these forms, and corresponding to them are therefore $gd$ modes. Thus, allowing for the strong coupling effects decreases the number of phonon branches from 9 to 5. The modes $u_{32}$, $v_{32}$, $u_{33}$, and $v_{33}$, which correspond in the weak coupling approximation to orbital waves, become nonphonon modes when the strong coupling effects are taken into account.

Expression (44) at $\nu \neq 0$ describes the system in a magnetic field. In the weak coupling approximation the number of phonon modes decreases from 9 to 6, and the variables $u_{13} + v_{23}$, $u_{33}$, and $v_{33}$ become nonphonon because of the appearance of the gap $-\mu H$ in the spectrum. In the system with strong coupling mode that becomes non-phonon upon application of a magnetic field is $u_{13} + v_{22}$ (the modes $u_{33}$ and $v_{33}$ in the case of strong coupling remain $gd$ ones also at $\nu = 0$), and the number of $gd$ modes decreases from 5 to 4.

To gain an idea of the total Bose spectrum (including the $gd$-branches) when strong coupling effects are taken into account, we write (44) at $H = 0$ ($\nu = 0$) in the form

$$\delta^2 F = 4(a + b + d)(u_{11} + v_{21})^2 + 4(c + e)[(u_{11} - v_{21})^2$$

$$+ (u_{21} + v_{11})^2] + 2(c - b - d)[(u_{13} - v_{23})^2 + (u_{12} - v_{22})^2$$

$$+ (u_{22} + v_{12})^2 + (u_{23} + v_{13})^2] - 4d(u_{23} - v_{13})^2 \tag{46}$$

$$+ (u_{22} - v_{12})^2] - 4(b + d)(u_{32}^2 + v_{32}^2 + u_{33}^2 + v_{33}^2)$$

For comparison, we write down $\delta^2 F$ in the weak coupling approximation, putting in (46) $a = b = c = -d = -2e = 1$:

$$\delta^2 F = 4[(u_{11} + v_{21})^2 + (u_{23} + v_{13})^2 + (u_{22} - v_{12})^2]$$

$$+ 2[(u_{11} - v_{21})^2 + (u_{21} + v_{11})^2 + (u_{13} - v_{23})^2 + (u_{12} - v_{22})^2 \tag{47}$$

$$+ (u_{22} + v_{12})^2 + (u_{23} + v_{13})^2] + 0[u_{32}^2 + u_{33}^2 + v_{32}^2 + v_{33}^2].$$

The form (47) has three eigenvalues 4, corresponding to the variables $u_{11} + v_{21}$, $u_{22} - v_{12}$, and $u_{23} - v_{13}$. It is just to these variables that the branches $E_1$ correspond as $T \to 0$. The other nonzero eigenvalue 2 correspond to six variables: $u_{21} + v_{11}$, $u_{12} - v_{22}$, $u_{13} - v_{23}$, $u_{11} - v_{21}$, $u_{23} + v_{13}$, and six $E_2$ branches as $T \to 0$.

The calculation of the Bose spectrum in Ref. 5 yields 6 $cl$-modes and three $2\Delta_0$ modes, i.e., as many as in the weak coupling case considered here. Formula (46) shows that in the general case allowance for the strong coupling effects leads to splitting. The $cl$-modes break up into two groups—two branches corresponds to the eigenvalue $4(c + e)$ and four correspond to the number $2(c - b - d)$. The three $2\Delta_0$ branches also break up into one branch with eigenvalue $4(a + b + d)$ and two branches with eigenvalue $-4d$. We note that no conclusion can be drawn from the data of Ref. 5 concerning the splitting of the branches.

The branches $u_{32}$, $u_{33}$, $v_{32}$, and $v_{33}$, which in the weak coupling approximation are orbital waves, go over into the normal flapping mode and the super-flapping mode when account is taken of the strong coupling effects, as shown by comparison with the data of Ref. 5.

Volovik[6] shown first, that the fermions in the He$^3$-A are chiral and a field theory in superfluid $^3$He-A which describes the dynamics of chiral fermion excitations that are interacting with the order parameter collective boson modes is similar to the theory of the electroweak interaction. The roles of photons and $W$ bosons are played by orbital waves and four quasi-Goldstone spin-orbit modes, which we obtained above, respectively.

An equation of the Dirac type for fermions in $^3$He-A near the poles of the Fermi sphere can be derived from the Bogoliubov equation, in which it is necessary to take into account the fluctuations $\delta A_{ai}$ of the order parameter $A_{ai}$ around its equilibrium value

$$A_{i\alpha}^{(0)} = \Delta_0 d_\alpha (e_1^i + i e_2^i).$$

Only certain combinations of fluctuations act on the fermions. These combinations form a "photon" field and $W$ field[6]:

$$A_1 + iA_2 = -\frac{k_F}{\Delta_0} d_\alpha 1^i \delta A_{i\alpha},$$

$$W_1^\alpha + iW_2^\alpha = \frac{k_F}{i\Delta_0} e^{\alpha\beta\gamma} d_\beta 1^i \delta A_{i\gamma},$$

$$A_3 = \delta k_F, \quad W_3^\alpha = \frac{1}{2k_F} e^{\alpha\beta\gamma} d^\beta \frac{\partial}{\partial t} d^\gamma;$$

$$A_0 = k_F \vec{1} \cdot \vec{v}_s, \quad W_0^\alpha = \frac{1}{2} k_F e^{\alpha\beta\gamma} d_\beta (\vec{1} \cdot \vec{\nabla}) d_\gamma.$$

(48)

# The Theory of the Collective Excitations in the A-Phase of He³

Equations (48) also incorporate the effect of the fluctuational spin $\vec{S} \sim [\vec{d}\partial \vec{d}/\partial t]$, which accounts for a third component of the $W$ field, in precisely the same way as density fluctuations account for a third component, along $\vec{1}$, of the "photon" field $\vec{A}$.

In terms of the fields in (48), the Bogoliubov equation for the Bogoliubov spinor $\psi$ takes the following form near the poles of the Fermi sphere:

$$\left[ \left( i\frac{\partial}{\partial t} - eA_0 - e\sigma^\alpha W_0^\alpha \right) + e(c_\| 1^i \tau_3 + c_\perp (e_1^i \tau_1 + e_2^i \tau_2)) \left( \frac{1}{i}\nabla_i - eA_i - e\sigma^\alpha W_i^\alpha \right) \right] \psi = 0$$

Here $\tau_i$ and $\sigma^\alpha$ are the Pauli matrices corresponding to the Bogoliubov isospin and ordinary spin. This equation is reminiscent of the Dirac equation for massless chiral fermions in the Weinberg-Salem theory. The primary distinction is in the anisotropy of He³-A along the $\vec{1}$ and $\vec{d}$ axes. The velocity $c_\| = v_F$ along $\vec{1}$ is far higher than the transverse velocity $c_\perp = \Delta_0/k_F$, and we have $W^\alpha d^\alpha = 0$; i.e., there are no $Z$ bosons. The charge $e = \vec{k} \cdot \vec{1}/k_F$ takes on the values $+1$ and $-1$ for fermions near the upper and lower poles, respectively.

An important point is that in the weak coupling approximation (in which the Fermi spheres with different spin projections do not interact with each other) there is an additional $SO(3)$ symmetry, which combines the "photons" and the $W$ bosons in a single triplet (more precisely, a sextet, when we take into account the polarization of the collective modes). In this approximation, the $W$ bosons, like the $gd$ orbital waves (or "photons"), have no mass in consequence with results obtained above. But as Brusov and Popov showed first[4] (see above) four additional $gd$-modes become nonphonon if we turn on the strong coupling corrections. Consequently, and in contrast with the Weinberg-Salem theory, the Higgs phenomenon is not required for the appearance of a mass of the $W$ bosons in He³-A.

## IV. THE LINEAR ZEEMAN EFFECT FOR CLAPPING AND PAIRBREAKING MODES

Below we consider the influence of magnetic fields on the $CE$ spectrum in He³-A.[7]

In accordance with Nasten'ka and Brusov's idea[8] for investigation of $CE$ spectrum in the presence of a magnetic field we must take into account both the additional term in $S_h$ and distortion of order parameter. The latter one in our case is equal[9]

$$c_{ia}(p) = c\sqrt{\beta V}\,\delta_{po}(\delta_{a1}\alpha_+ + i\delta_{a2}\alpha_-)(\delta_{i1} + i\delta_{i2}). \tag{50}$$

Here $\alpha_\pm = \dfrac{\Delta_\uparrow \pm \Delta_\downarrow}{2\Delta}$, $\Delta^2_{\uparrow\downarrow} = N(0)(\tau \pm \eta h)/2\beta_{245}$,

$$\eta = (N(0)/N(0))T_c \ln(1.14\epsilon_0/T_c),\ h = \frac{\mu_0 H}{T_c},\ \Delta = 2cZ -$$

is a single fermion spectrum gap, determined by next gap equation

$$\frac{1}{g} = -\frac{Z^2}{\beta V}\frac{1}{(\alpha_+^2 + \alpha_-^2)}\sum_p \left[\frac{(\alpha_+ + \alpha_-)^2 \sin^2\theta}{\omega^2 + (\xi - \mu H)^2 + \Delta^2 \sin^2\theta(\alpha_+ + \alpha_-)^2}\right.$$
$$\left. + \frac{(\alpha_+ - \alpha_-)^2 \sin^2\theta}{\omega^2 + (\xi + \mu H)^2 + \Delta^2 \sin^2\theta(\alpha_+ - \alpha_-)^2}\right] \tag{51}$$

One has for $G^{-1}$

$$G^{-1} = \begin{pmatrix} Z^{-1}(i\omega - \xi + \mu H \sigma_3)I_+;\ 2c(n_1 + in_2)(\alpha_+\delta_1 + i\alpha_-\sigma_2)I_- \\ -2c(n_1 - in_2)(\alpha_+\xi_1 - i\alpha_-\sigma_2)I_-;\ Z^{-1}(-i\omega + \xi + \mu H\sigma_3)I_+ \end{pmatrix} \tag{52}$$

Inverting $G^{-1}$, one gets

$$G = \begin{pmatrix} G_{11} & G_{12} \\ G_{21} & G_{22} \end{pmatrix},$$

where

$$G_{11} = \begin{pmatrix} \dfrac{a^+ + b}{d_1} & \dfrac{a^+ - b}{d_2} \end{pmatrix} I_+,\ G_{12} = \begin{pmatrix} q_1 \mp q_2 & \dfrac{q_1 - iq_2}{d_1} \\ & d_2 & \end{pmatrix} I_-,$$

$$G_{21} = \begin{pmatrix} 0 & \dfrac{-q_1^+ + iq_2^+}{d_2} \\ \dfrac{-q_1^+ - iq_2^+}{d_1} & 0 \end{pmatrix} I_+,\ G_{22} = \begin{pmatrix} \dfrac{-a^+ + b}{d_2} & 0 \\ 0 & -a^+ - b \\ & & d_1 \end{pmatrix} I_-,$$

where $a = Z^{-1}(i\omega - \xi),\ b = Z^{-1}\mu H,\ q_1 = Z^{-1}\Delta(n_1 + in_2)\alpha_+,$

$q_2 = iZ^{-1}\Delta(n_1 + in_2)\alpha_-,\ d_{1,2} = Z^{-2}(\omega^2 + (\xi \mp \mu H)^2 + \Delta^2 \sin^2\theta(\alpha_+ \pm \alpha_-)^2)$

## The Theory of the Collective Excitations in the A-Phase of He³

Using the expression (52) for $G$ and next one for $u$

$$u = (\beta V)^{-1/2} \begin{pmatrix} 0, & (n_1 - n_{2i})c_{ia}(p_1 + p_2)\sigma_a \\ -(n_{1i} - n_{2i})c_{ia}(p_1 + p_2)\alpha_a, & 0 \end{pmatrix}$$

one could obtain next quadratic part of $S_h$

$$\begin{aligned} S_h = g^{-1} \sum_{p,i,a} c_{ia}^+ c_{ia} &+ \frac{Z^2}{\beta V} \sum_{p+p=p} n_{1i} n_{1j} \{ [-\Delta^2(n_1 + in_2)^2 \\ &\times (-\partial_3 c_{i3}^+ c_{j3}^+ + \partial_1(c_{i1}^+ c_{j1}^+ - c_{i2}^+ c_{j2}) + i\partial_2(c_{i1}^+ c_{j2}^+ + c_{i2} c_{j2}^+) \; h.c.] \\ &+ D_3(c_{i3}^+ c_{j3} + c_{i3} c_{j3}^+) \\ &+ iD_2(c_{i2}^+ c_{j1} + c_{i1} c_{j2}^+ - c_{i2} c_{j1}^+ - c_{i1}^+ c_{j2}) \}. \end{aligned} \tag{53}$$

Here

$$\begin{aligned} \partial_{1,2} &= \frac{(\alpha_+ + \alpha_-)^2}{d_1(1)d_1(2)} \pm \frac{(\alpha_+ - \alpha_-)^2}{d_2(1)d_2(2)} \\ D_{1,2} &= \frac{(a^+(1) + b)(a^+(2) + b)}{d_1(1)d_1(2)} \pm \frac{(a^+(1) - b)(a^+(2) - b)}{d_2(1)d_2(2)} \\ \partial_3 &= (\alpha_+^2 - \alpha_-^2)\left(\frac{1}{d_1(2)d_2(1)} + \frac{1}{d_1(1)d_2(2)}\right) \\ D_3 &= \frac{(a^+(1) + b)(a^+(2) - b)}{d_1(1)d_2(2)} + \frac{(a^+(1) - b)(a^+(2) + b)}{d_2(1)d_1(2)} \end{aligned} \tag{54}$$

By diagonalizing the quadratic form we get its canonical form (here $u_{ia} = \text{Re } c_{ia}$, $v_{ia} = \text{Im } c_{ia}$ and $\Sigma$ means $\Sigma_{p_1+p_2}$):

$$\begin{aligned} S_h = &\left[g^{-1} + \frac{2Z^2}{\beta V} \sum D_3 \cos^2\theta\right](u_{33}^2 + v_{33}^2) \\ &+ \left[g^{-1} + \frac{2Z^2}{\beta V} \sum (D_1 + D_2)\cos^2\theta\right]((u_{31} + v_{32})^2 + (u_{32} + v_{31})^2) \\ &+ \left[g^{-1} + \frac{2Z^2}{\beta V} \sum (D_1 - D_2)\cos^2\theta\right]((u_{31} - v_{32})^2 + (u_{32} - v_{31})^2) \end{aligned}$$

$$+ \left[g^{-1} + \frac{Z^2}{\beta V} \sum D_3 \sin^2\theta\right]((v_{31} + u_{23})^2 + (u_{13} - v_{23})^2)$$

$$+ \left[g^{-1} + \frac{Z^2}{\beta V} \sum (-\Delta^2 \sin^2\theta \partial_3 + D_3)\sin^2\theta\right](u_{13} + v_{23})^2$$

$$+ \left[g^{-1} + \frac{Z^2}{\beta V} \sum (\Delta^2 \sin^2\theta \partial_3 + D_3)\sin^2\theta\right](v_{13} - u_{23})^2 \quad (55)$$

$$+ \left[g^{-1} + \frac{Z^2}{\beta V} \sum (D_1 + D_2)\sin^2\theta\right]$$

$$\times ((u_{12} + v_{11} + v_{21} + v_{21})^2 + (u_{11} + v_{12} - u_{22} - v_{21})^2)$$

$$+ \left[g^{-1} + \frac{Z^2}{\beta V} \sum (D_1 - D_2)\sin^2\theta\right]((u_{12} - v_{11} - u_{21} + v_{22})^2$$

$$+ (u_{11} - v_{12} + u_{22} - v_{21})^2)$$

$$+ \left[g^{-1} + \frac{Z^2}{\beta V} \sum \sin^2\theta(D_1 + D_2 - \Delta^2 \sin^2\theta(\partial_1 + \partial_2))\right]$$

$$\times (u_{12} + v_{11} - u_{21} - v_{22})^2$$

$$= + \left[g^{-1} + \frac{Z^2}{\beta V} \sum \sin^2\theta(D_1 - D_2 - \Delta^2 \sin^2\theta(\partial_1 - \partial_2))\right]$$

$$\times (u_{11} - v_{11} + u_{21} - u_{22})^2$$

$$+ \left[g^{-1} + \frac{Z^2}{\beta V} \sum \sin^2\theta(D_1 + D_2 + \Delta^2 \sin^2\theta(\partial_1 + \partial_2))\right]$$

$$\times (u_{11} + v_{12} + u_{22} + v_{21})^2$$

$$+ \left[g^{-1} + \frac{Z^2}{\beta V} \sum \sin^2\theta(D_1 - D_2 + \Delta^2 \sin^2\theta(\partial_1 - \partial_2))\right]$$

$$\times (u_{11} - v_{12} - u_{22} + v_{21})^2.$$

Equation $\det Q = 0$, where $Q$ is the matrix of quadratic form (55), gives us 18 equations which completely determine 18 collective modes in $He^3$-A in arbitrary magnetic field and arbitrary $CE$ momenta.

## The Theory of the Collective Excitations in the A-Phase of He³

Below we consider the case of small $\vec{H}$ and $\vec{k} = 0$, and calculate the linear correction to the CE spectrum. Putting $\vec{k} = 0$ and retaining the terms up to first order on field, we obtained next equations

I) $\quad \displaystyle\int_0^\pi (1 - (1 + 2c)I)\cos^2\theta\sin\theta d\theta = 0; \quad u_{33}, \ v_{33}$

$\displaystyle\int_0^\pi (-1 + (1 + 2c)I)\cos^2\theta\sin\theta d\theta$

$\pm \ \gamma H \displaystyle\int_0^\pi \frac{4c}{1 + 4c}(1 + 2cI)\cos^2\theta\sin_3\theta d\theta = 0;$

$u_{31} + u_{32}, \ u_{32} + v_{31}(u_{31} - v_{32}, \ u_{32} - v_{31})$

II) $\quad \displaystyle\int_0^\pi I\sin^3\theta d\theta = 0; \quad u_{23} - v_{13}$

$\displaystyle\int_0^\pi I\sin^3\theta d\theta \pm \gamma H \displaystyle\int_0^\pi \frac{4c}{1 + 4c}(2 - I)\sin^3\theta d\theta = 0;$

$u_{11} + v_{12} + u_{22} + v_{21} \ (u_{11} - v_{12} - u_{22} + v_{21})$

III) $\quad \displaystyle\int_0^\pi (1 + 2c)I\sin^3\theta d\theta = 0; \quad v_{13} + u_{23}, \ u_{13} - v_{23}$

$\displaystyle\int_0^\pi (1 + 2c)I\sin^3\theta d\theta \pm \gamma H \displaystyle\int_0^\pi \frac{4c}{1 + 4c}(1 + 2cI)\sin^3\theta d\theta = 0;$

$u_{12} + v_{11} + u_{21} + v_{22}, \ u_{11} + v_{12} - u_{22}$

$- \ v_{21}(u_{12} - v_{11} - u_{21} + v_{22}, \ u_{11} - v_{12} + u_{22} - v_{21})$

IV) $\quad \displaystyle\int_0^\pi (1 + 4c)I\sin^3\theta d\theta = 0; \quad u_{13} + v_{23}$

$\displaystyle\int_0^\pi (1 + 4c)I\sin^3\theta d\theta \pm \gamma H \displaystyle\int_0^\pi 4cI\sin^3\theta d\theta = 0;$

$u_{12} + v_{11} - u_{21} - v_{22} \ (u_{12} - v_{11} + u_{21} - v_{22})$

where $I = \dfrac{1}{\sqrt{1 + 4c}} \ln \dfrac{\sqrt{1 + 4c} + 1}{\sqrt{1 + 4c} - 1}, \quad c = \dfrac{\Delta^2\alpha_+^2\sin^2\theta}{\omega^2}; \ \gamma H = \dfrac{\alpha^-}{\alpha^+}$. We have 4 groups of equations each with 3 or 6 equations. The I and II groups of equations describe $gd$-modes. For these modes it's needed to take into account the quadratic on field corrections. In Ref. 4 (see part 3 above) the conclusion has been made that in the presence of magnetic field 3 from 9 $gd$-modes become nonephonon because of the appearance of the gap $\sim \mu H$ in their spectrum.

The III and IV groups of equations describe the $cl$ and $pb$-modes. If we write these equations as $F_0(E) \pm \gamma H F_1(E) = 0$ and will search $E$ as $E = E_0 \pm \gamma H E_1$, we obtain

$$E = E_0 \pm \gamma H E_1 = E_0 \left( 1 \mp \frac{F_1(E_0)}{\frac{8}{3}} - F_1(E_0) \right).$$

As we noted above the values of $E_0$ for $cl$ and $pb$-modes were obtained by Brusov and Popor earlier.[2] Using these data we'll obtained for the energies of $cl$ and $pb$-modes:

$cl$: $E_1 = (1.17 - i0.13)\Delta_0$

$E_{2,3} = (1.17 \pm 1.62\gamma H)\Delta_0 - i(0.13 \pm 1.33\gamma H)\Delta_0$

$pb$: $E_1 = (1.96 - i0.31)\Delta_0$

$E_{2,3} = (1.96 \pm 1.18\gamma H)\Delta_0 - i(0.31 \pm 0.69\gamma H)\Delta_0$

So for small $\vec{H}$ we have 3-fold splitting of the $cl$- and $pb$-modes; i.e. we got a linear Zeeman effect for these modes. Magnetic fields lift completely the degeneracy of $pb$-modes, and particularly lift the degeneracy of $cl$-modes, each branch of latter modes remain twice degenercity.

Note that magnetic field changes both the real parts of $CE$ energies and the imaginary ones i.e. changes $CE$ frequencies and damping, even in linear approximation. The damping of some of the modes increases while other ones decreases with a magnetic field.

We could compare our results with ones from Refs. 10 and 11. Note that our $CE$ spectrum from one side is different from one in Ref. 10 and 11 via the existence of additional 4 quasi $gd$-modes. From other side our results for $CE$ energies are more exact because we take into account the $CE$ damping via the processes of Cooper pair breaking.

In Ref. 10 some modes remain unchanged while frequencies of some others are shifted from $\omega_i$ to $(\omega_i^2 + \Omega^2)^{1/2}$ (here $\Omega$ is the effective Larmor frequency). Such quadratic on field shifts are obtained without taking into account the gap distortion. Our results for real parts of $CE$ energies are mored close to ones of Ref. 11. (In Ref. 11, however the $CE$ damping does not take into account and influence of magnetic field on it does not investigate.) There one has the linear splitting of some $CE$ frequencies

$$\omega_{i\sigma}\left(\frac{T}{T_{c\sigma}}\right) = \frac{T_{c\sigma}}{T_{co}} \omega_i\left(\frac{T}{T_{c\sigma}}\right), \quad i = \mathit{nfl}, \mathit{cl}, \mathit{sfl}, \mathit{scl}, \sigma = \uparrow, \downarrow$$

via gap distortion while frequencies of other modes are shifted from the zero-field values $\omega_i$ to $(\omega_i^2 + \Omega^2)^{1/2}$, i.e. retain unchanged in linear in field approximation.

Other conclusion which has been made by us is that the linear Zeeman effect for *cl*- and *pb*-modes takes place via the distortion of order parameter only (or particle-hole asymmetry) for case of zero $\vec{k}$. For nonzero $\vec{k}$ there is in principle a possibility that additional term in the action (without gap distortion) will lead to the linear on field corrections to mode energies.

Mentioned above 3-fold splitting of the *cl*- and *pb*-modes could be observed in zero sound experiments. Note that in case of *A*-phase in opposite to case of *B*-phase gap distortion is linear in field that leads to the possibility of its effect observation in moderate fields.

I thank very much M.Y. Nasten'ka for supporting me during my stay at Northwestern. I would also like to thank T.V. Filatova-Novoselova for five years of collaboration. I thank V.N. Popov and M.V. Lomakov for taking part in the investigation of *CE* in He$^3$-*A*, G.E. Volovik, J.B. Ketterson, W.P. Halperin, J.A. Sauls, and Z. Tešanovic for helpful discussions, and Northwestern University, I. Weil and IREX for their hospitality. At last I thank very much Z. Tešanović for his hospitality during my stay at Johns Hopkins University.

## REFERENCES

1. V. Alonso and V. N. Popov, *Zh. Eksp. Teor. Fiz.* **73**, 1445 (1977), [*Sov. Phys. JETP* **46**, 760 (1977)].
2. P. N. Brusov and V. N. Popov, "The superfluidity and collective properties of quantum liquids," "Nauka," Moskow, 1988, p.p. 215.
3. G. E. Volovik and M. V. Khazan, *Zh. Eksp. Teor. Fiz.* **85**, 948 (1983) [*Sov. Phys. JETP* **58**, 551 (1983)].
4. P. N. Brusov and V. N. Popov, *Zh. Eksp. Teor. Fiz.* **79**, 1871 (1980) [*Sov. Phys. JETP*, **52**, 945 (1980)].
5. P. Wölfe, *Physica* **90B**, 96 (1977).
6. G. E. Volovik, *Pis'ma Zh. Eksp Teor. Fiz.* **43**, 535 (1986), [*JETP Lett.* **43**, 693 (1986)].
7. P. N. Brusov, M. Y. Nastenka, T. V. Filatova-Novoselova, M. V. Lomakov, and V. N. Popov, Preprint of Northwestern University, 1989, submitted to Physical Review Letters.
8. M. Y. Nasten'ka and P. N. Brusov, *Phys. Lett.* (to be published).
9. G. E. Gurgenishvili and G. A. Kharadze, "Superfluid phases of liquid He$^3$," Martseba, Tbilisi, 1986, pp. 180.
10. L. Tewordt and N. Schopohl, *J. Low Temp. Phys.* **34**, 489 (1979).
11. N. Schopohl, W. Marquardt, L. Tewordt, ibid, **59**, 469 (1985).

Yi-Hong Chen and Frank Wilczek
Institute for Advanced Study
Princeton, NJ 08540

# Induced Quantum Numbers and Surface Currents in Some 2+1 Dimensional Models

The fermion current induced by slow variations in background scalar and gauge fields are computed for a class of 2+1 dimensional $\sigma$-like models. Local current densities proportional to topological currents in the background fields are found. The coefficient depends discontinuously on certain field ratios. The discontinuity and the conservation of fermion currents implies the existence of surface correction currents, which have the form of London supercurrents. A general treatment of such correction currents is included. The induced fermion numbers we find, mesh nicely with recent results on induced angular momentum and induced statistics. In particular, the spin and statistics is intimately related to the global parity anomaly. Lattice realizations are suggested.

## I. INTRODUCTION

Recently there has been renewed interest in 2+1 dimensional physics, and especially the description of magnetic systems. This interest has been largely stimulated by the properties of various materials containing $CuO_2$ layers, which are quite remarkable and anomalous even apart from their association in certain cases with high-temperature superconductivity.

Many aspects of the theory of these materials are apparently confused and certainly controversial. There are two essential problems: the identification of an appropriate idealized model, and the analysis of the model once chosen. Given present uncertainties, it may be wise to explore qualitative features of possible relevant field theory models in some generality, so that we may "know what to look for." This strategy also has the advantage that the produced models may apply to other physical systems, even if not to $CuO_2$ layers.

In this spirit we analyze here some interesting qualitative features of a class of model quantum field theories in 2+1 dimensions. Let us hasten to add that our choice of models is far from random. They commonly feature scalar fields, with internal degrees of freedom, interacting with fermions. These are meant to suggest, among other things, the effective interaction of a magnetic order parameter—magnetization or staggered magnetization—with electrons. Some connections with various specific speculations in the literature, and the possibility of concrete lattice realizations of the models will appear in the following.

## II. PRELIMINARIES, BASIC MODEL

It seems appropriate to review briefly some properties of fermions in 2+1 dimensions, that may be unfamiliar and are rather peculiar.

For the Dirac matrices $\Gamma_0$, $\Gamma_1$, $\Gamma_2$ we may take simply $(\sigma_2, i\sigma_1, i\sigma_3)$, where the $\sigma$'s are the usual Pauli matrices, since they obey the appropriate anticommutation relations. Notice that these $\Gamma_i$ are all pure imaginary. Thus it would be consistent to require $\Psi$ to be real; however in the models of interest to us we will be working with complex $\Psi$. The Lagrangian for a free fermion

$$L = \overline{\Psi} i \Gamma_\mu \partial_\mu \Psi \qquad (2.1)$$

is invariant under the discrete parity, time-reversal, and charge conjugation transformations, under which the fields transform as

$$P : \Psi(t, x_1, x_2) \to \sigma_1 \Psi(t, -x_1, x_2) = i\Gamma_1 \Psi(t, -x_1, x_2) \qquad (2.2)$$

$$T : \Psi(t, \vec{x}) \to i\sigma_2 \Psi^\dagger(-t, \vec{x}) = i\Gamma_0 \Psi^\dagger(-t, \vec{x}) \qquad (2.3)$$

$$C : \Psi \to \Psi^\dagger \qquad (2.4)$$

As usual, $P$ and $C$ are implemented by unitary, and $T$ by an antiunitary, transformation $\Psi \to U\Psi U^\dagger$. Note that parity involves reflection about one axis; simultaneous inversion about both spatial axes is just rotation through $\pi$. If $\Psi$ is real, it is unchanged under $C$ transformations and remains real under $P$ and $T$ transformations.

An important point is that the mass term

$$\Delta L = m\overline{\Psi}\Psi$$

is *odd* under $P$ and $T$.

We shall be considering models in 2+1 dimensions containing a doublet of two-component fermions and a singlet and triplet of pseudoscalar fields $\eta$, $\vec{\phi}$. The interaction Lagrangian is

$$L_{int} = \eta\overline{\Psi}\Psi + \overline{\Psi}\vec{\tau}\cdot\vec{\phi}\Psi + V(\eta, \vec{\phi}) \tag{2.5}$$

where $V(\eta, \vec{\phi})$ contains scalar self-interactions. This interaction is reminiscent of that in the original Gell-Mann-Levy σ-model.[1] We shall also consider models that contain gauge fields, such that the $U(1)$ and $SU(2)$ symmetries of (2.5) become local symmetries.

## III. INDUCED CURRENTS

We now consider the flow of fermion currents in response to slow space-time variation of $\eta$ and $\vec{\phi}$, due to the interaction (2.5). As is familiar from other contexts, consideration of such flows can be useful in understanding the spin and statistics of solitons, among other things.

It is straightforward to calculate the induced current, following[2] (see Appendix). The result is

$$\langle j^\mu \rangle \equiv \langle \overline{\Psi}\Gamma^\mu\Psi \rangle = \frac{1}{8\pi}\Theta(|\vec{\phi}| - |\eta|)\epsilon^{\mu\beta\gamma}\epsilon_{abc}\hat{\phi}^a\partial_\beta\hat{\phi}^b\partial_\gamma\hat{\phi}^c \tag{3.1}$$

where $\hat{\phi}^a = \phi^a/|\vec{\phi}|$, $\Theta$ is the function equal to 1, $\frac{1}{2}$, or 0, depending on whether its argument is positive, zero, or negative.

The interpretation of this result requires some comment. Let us assume, first, that $\eta = 0$. Then the form of $\langle j^\mu \rangle$ is entirely reminiscent of the corresponding results in 3+1 and 1+1 dimensions. Here, as in those cases, the induced current is proportional to the density of an identically conserved current, and the associated charge has a topological interpretation (see below). The form of the current for $\eta \neq 0$, on the other hand, is rather surprising. After all, $\langle \eta \rangle$ contributes to the mass of the fermions. So what we have found is that as we vary the mass the current does not change at all until a critical value of the fermion mass is reached, at which point the current suddenly drops to zero. This behavior is quite different from the smooth behavior found in other dimensions.

How can such a discontinuity arise, in what seems to be a totally non-singular context? The expression (3.1) for the current was derived on the assumption of slow space-time variation—and the criterion of "slowness" is that the $\eta$ and $\vec{\phi}$ fields vary little, as a fraction of their value, over the Compton wavelength of the fermions. But the magnitudes of the masses of the two fermions are $|\eta \pm \phi_3|$. Thus as either of these quantities goes to zero, the criterion for slow variation becomes harder and harder to meet. In other words, the discontinuity arises from an interchange of limits, and would presumably be smoothed out in a calculation to higher order in gradients. Nevertheless, it is true as a limiting case and is quite significant physically as we shall soon see.

It is of interest also to compute the current in the more general model, where the $U(1)$ fermion number and $SU(2)$ isospin symmetries of the model are gauged. The result is (see Appendix)

$$\langle j^\mu \rangle = \frac{1}{8\pi} \Theta(|\vec{\phi}| - |\eta|)(\epsilon^{\mu\beta\gamma}\epsilon_{abc}\hat{\phi}^a D_\beta \hat{\phi}^b D_\gamma \hat{\phi}^c - g\epsilon^{\mu\beta\gamma}F^c_{\beta\gamma}\hat{\phi}^c)$$

$$+ \frac{e^2}{8\pi} \frac{\eta}{|\eta|} \Theta(|\eta| - |\vec{\phi}|)\epsilon^{\mu\beta\gamma}F_{\beta\gamma} \qquad (3.2)$$

where

$$D_\beta \phi^i = \partial_\mu \phi^i + g\epsilon^{ijk}A^j_\mu \phi^k \qquad (3.3)$$

is the covariant derivative, $e$ is the $U(1)$ coupling constant, and $F^c_{\beta\gamma}$ and $F_{\beta\gamma}$ are respectively the $SU(2)$, $U(1)$ field strengths. The term containing $F^c_{\beta\gamma}$ is necessary to ensure conservation of $\langle j^\mu \rangle$. This term, and the final term which is its abelian analogue (and has precisely the same graphical origin), form what is generally called the "parity anomaly." Equation (3.2) presents a remarkably balanced appearance.

## IV. RESULTS

The induced current may be used to calculate the induced fermion numbers of scalar field configurations, assuming that the characteristic length scale of the latter extends over many Compton wavelengths of the fermion. In complete generality, it may be used to calculate the fractional part of these quantum numbers. Two cases of interest, are the baby Skyrmion and, in the gauged model, the $SU(2)$ vortex.

We find, for the baby Skyrmion

$$Q_s = \int \langle j^0 \rangle d^2x = 1 \qquad (4.1)$$

while for the $SU(2)$ vortex

$$Q_v = \int \langle j^0 \rangle d^2x = \frac{1}{2} \qquad (4.2)$$

assuming, respectively, $|\vec{\phi}| > |\eta|$ everywhere, or at spatial infinity. Some of the calculational details are presented in the Appendix.

These results make contact with some recent suggestions in the condensed matter physics literature. In particular, baby Skyrmions carrying unit fermion number should plausibly be quantized as fermions and form an isodoublet. If we identify isospin in our model with the ordinary spin, the scalar fields as magnetic order parameter, and the interaction of our fermion fields with the scalar background as arising from some filled band whose gap is dominated by the alignment of spin with respect to the order parameter, then we have shown that the solitons, that are electrically neutral, nevertheless carry spin. This is a model for Anderson's "spinons"[3], or for the objects conjectured to exist in insulating paramagnets by DPW.[4] It must be emphasized, however, that the spin and statistics of these solitons depends completely upon the coefficient of the so-called Hopf term.[5] The assignment above is appropriate to zero bare coefficient for this term, that is to zero coefficient before the heavy fermion is integrated out. The bare coefficient, however, is sensitive to the ultraviolet (short-distance) behavior of the theory, as of course the fact that the coefficient can be affected by arbitrarily heavy fermions exemplifies. Also, the Hopf term, like the vector potential in the Aharonov-Bohm effect[6]—to which in fact it is deeply related—has no classical analogue. Therefore, its coefficient cannot be deduced from the long-distance or semiclassical behavior of an underlying microscopic model, but can only be deduced from a calculation where short-distance fluctuations, both thermal and quantum, are fully taken into account.

Also, the "half fermion" quantum number for the vortex is strongly reminiscent of the quasiparticles proposed by Laughlin.[7] One of us has argued elsewhere that these quantum numbers arise naturally for vortices in an *abelian* gauge model.[8]

In both these cases, we can restrict and classify the possibilities considerably by demanding that time-reversal symmetry be respected, or that it be broken spontaneously in a structured way.[9]

Remarkably, for $SU(2)$ vortices there are strong *a priori* restrictions on the allowed spin and statistics. These restrictions can be thought of as arising in either of two ways, that seem very different but as we shall see lead to the same conclusions. First, let us recognize that unlike abelian vortices, that carry on ordinary additive quantum number (their flux), $SU(2)$ vortices carry only a $Z_2$ flux. Thus two vortices can annihilate into a topologically trivial configuration, that should have a sensible interpretation in terms of the ordinary particle content of the theory. We have seen that each vortex carries fermion number $\frac{1}{2}$, so that the pair carries fermion number 1. In the topologically trivial sector of the theory, of course, we know that particles carrying fermion number 1 are ... fermions. Now the statistics of a pair of particles each with statistical parameter θ is 4θ. So we must demand

$$4\theta = \pi, \ modulo \ 2\pi \tag{4.3}$$

or, for the statistics (and spin) of the vortex

$$\theta = \pi/4, \ 3\pi/4, \ 5\pi/4, \ or \ 7\pi/4 \tag{4.4}$$

The second way, is to realize again that the statistics of the vortex is governed by the value of the coefficient of an appropriate Chern-Simons term, this time the non-abelian one.[10] Now the non-abelian Chern-Simons term is only properly globally gauge invariant when its coefficient takes on certain quantized values. A simple calculation shows, that these values lead precisely to the possibilities listed above for vortex statistics. In our opinion, the concordance of these methods is quite pretty, and gives an unusually concrete physical interpretation of a global anomaly.

## V. SURFACE CURRENTS

The most novel feature of the current (3.2), compared with previous results in other dimensions, is the appearance of discontinuities. Of course the current flow cannot really just stop abruptly at a surface, since this would be inconsistent with its conservation. To determine the form of the correction, let us consider a general current of the form

$$\tilde{j}^\mu = k_\mu f(\vec{r}) \tag{5.1}$$

where

$$\partial_\mu k^\mu = 0 \tag{5.2}$$

and $f(\vec{r})$ is a general function of position that has

$$k^\mu \partial_\mu f(\vec{r}) \neq 0 \tag{5.3}$$

As it stands, this current is not conserved. However, we can largely determine the possible origin and form of a local current that would compensate for this non-conservation. Indeed, consider the Lagrangian

$$L = \tilde{j}^\mu A_\mu - k^\mu \partial_\mu \phi f(\vec{r}) + \frac{1}{2}(\partial_\mu \phi - A_\mu)^2 \qquad (5.4)$$

where $\phi$ is a scalar field, and $A_\mu$ a gauge field such that the Lagrangian is invariant under $\phi \rightarrow \phi + \chi$, $A_\mu \rightarrow A_\mu + \partial_\mu \chi$. The equation of motion for the scalar field is

$$\partial^\mu(\partial_\mu \phi - A_\mu) = k^\mu \partial_\mu f(\vec{r}) \qquad (5.5)$$

It is easy to see that the current $j_\mu = \dfrac{\partial L}{\partial A_\mu} = \tilde{j}_\mu - (\partial_\mu \phi - A_\mu)$ is now conserved.

In the case where $f(\vec{r}) = \Theta(x)$, the step function, the correction current lives on $x = 0$, the singularity of the $\Theta$ function. The physical interpretation of this surface current is as follows. The jump in the $\Theta$ function occurs where some fermion mass goes through zero. The mass will have non-zero magnitude on either side, and we can anticipate there to be fermion zero modes that live on the wall, in the sense that for fermions in these modes motion tangent to the wall costs little energy. It is motion in these modes that builds up the surface current.[11]

If we fix the gauge such that $\phi = 0$, then the correction current $\Delta j_\mu = A_\mu$ has the form of the London current. That is, the surface current has the same form as the diamagnetic current that produces the Meissner effect in superconductors. In this gauge, the $\phi$ field equation, i.e. (5.5) with $\phi = 0$ becomes a constraint.

## VI. LATTICE FORMULATION

The field theory models discussed above can also be put on a lattice. This is important, in two respects. First, the lattice models may have a direct use as models for crystalline solids. Second, one may wonder whether the effects we have discussed above, which are so closely tied to topology, survive latticization. Concretely, one may well be suspicious that some analogue of the fermion doubling phenomenon removes the possibility of inducing interesting quantum numbers by fermion vacuum polarization.

At one level this sort of suspicion is alleviated by the beautiful work of Frohlich and Marchetti.[12] These authors have constructed anyon lattice field theories in full rigor. Thus there is little doubt that the effective theories we are aiming towards are consistent; and we need only wonder whether they result from lattice fermion theories, upon integrating out the fermions. This, too, we believe is not a serious difficulty. The point is that 2-component fermions in 2+1 dimensions, unlike chiral fermions in 3+1 dimensions, support mass-like terms. Thus we can avoid the doubling problem, without spoiling any crucial symmetry, by using a version of Wilson fermions.[13]

Let us quickly review the construction of a lattice formulation. Consider the Lagrangian

$$L = \overline{\Psi}\partial\!\!\!/\Psi + \overline{\Psi}(\vec{\phi}\cdot\vec{\tau} + \eta)\Psi + \lambda a \overline{\Psi}\nabla^2\Psi + (\partial_\mu\vec{\phi})^2 + (\partial_\mu\eta)^2 + V(\eta, \vec{\phi}) \quad (6.1)$$

where $a$ is the spacing of the fermion lattice. The third term is added to eliminate the fermion doubling problem, it goes to zero in the continuum limit ($a \to 0$).

There are many ways in which one can put the scalar fields on a lattice. Here we choose, for the $\vec{\phi}$ fields, a square lattice interlaced with the fermion lattice. We will put the $\eta$ field on the sites of the fermion lattice.

For slowly varying $\vec{\phi}$ and $\eta$, the term $\overline{\Psi}(\vec{\phi}\cdot\vec{\tau} + \eta)\Psi$ becomes effectively a mass term. The Dirac equation for the fermions is then

$$(\partial\!\!\!/ + m + \lambda a \overline{\Psi}\nabla^2)\Psi = 0 \quad (6.2)$$

On the lattice it becomes

$$\Gamma_0 \dot{\Psi}_{mx,my} = \Gamma_1 \frac{1}{2a}(\Psi_{mx+1,my} - \Psi_{mx-1,my}) + \Gamma_2 \frac{1}{2a}(\Psi_{mx,my+1} - \Psi_{mx,my-1})$$

$$+ m\Psi_{mx,my} + \frac{\lambda}{a}(\Psi_{mx+1,my} + \Psi_{mx-1,my} - 2\Psi_{mx,my})$$

$$+ \frac{\lambda}{a}(\Psi_{mx,my+1} + \Psi_{mx,my-1} - 2\Psi_{mx,my}) \quad (6.3)$$

For plane wave solution $\Psi \sim e^{i\vec{k}\cdot\vec{r} - iEt}$ the energy spectrum is

$$E^2 = \frac{\sin^2(k_x a)}{a^2} + \frac{\sin^2(k_y a)}{a^2}$$

$$+ \left[m + \frac{4\lambda}{a}\left(\sin^2\left(\frac{k_x a}{2}\right) + \sin^2\left(\frac{k_y a}{2}\right)\right)\right]^2 \quad (6.4)$$

From this one can see that the possible fermion doubling at $k_x a = \pi$, $k_y a = \pi$ is eliminated due to the presence of the last term.

# APPENDIX: SOME DETAILS OF CALCULATION

When the $U(1)$ and $SU(2)$ symmetries are not gauged, the induced current must have the general form

$$\langle j^\mu \rangle \equiv \langle \overline{\Psi} \Gamma^\mu \Psi \rangle = C \epsilon^{\mu\beta\gamma} \epsilon_{abc} \langle \hat{\phi}^a \partial_\beta \hat{\phi}^b \partial_\gamma \hat{\phi}^c \rangle \tag{.1}$$

To determine $C$, we calculate $j^\mu$ for a configuration of $\hat{\phi}$ where $\phi_3 \sim 1$, $\phi_2$ and $\phi_1 \ll 1$. The result for a general configuration follows by symmetry.

Then

$$\langle j^\mu \rangle \equiv \langle \overline{\Psi} \Gamma^\mu \Psi \rangle$$

$$= C \epsilon^{\mu\beta\gamma} \epsilon_{123} \langle \hat{\phi}^3 [\partial_\beta \hat{\phi}^1 \partial_\gamma \hat{\phi}^2 - \partial_\beta \hat{\phi}^2 \partial_\gamma \hat{\phi}^1] \rangle$$

$$= 2C \epsilon^{\mu\beta\gamma} p_\beta^1 p_\gamma^2 \tag{.2}$$

where $p_\beta^i \equiv \langle \partial_\beta \hat{\phi}^i \rangle$.

Consider the lowest order Feynman graphs that contribute to $j^\mu$ (Fig. 1)

$$j^\mu = 2 \int \frac{d^3\vec{k}}{(2\pi)^3} Tr\left[\Gamma^\mu \frac{1}{(\slashed{k} - \slashed{p}_1) - \Delta} \tau_1 \frac{1}{\slashed{k} - \Delta} \tau_2 \frac{1}{(\slashed{k} + \slashed{p}_2) - \Delta}\right] \tag{.3a}$$

$$= -2 \int \frac{d^3\vec{k}}{(2\pi)^3} Tr\left[\Gamma^\mu \frac{i\tau_3(\slashed{k} - \slashed{p}_1 + \Delta)(\slashed{k} + \tilde{\Delta})(\slashed{k} + \slashed{p}_2 + \Delta)}{((k - p_1)^2 + \Delta^2)(k^2 + \tilde{\Delta}^2)((k + p_2)^2 + \Delta^2)}\right] \tag{.3b}$$

where $\Delta = |\vec{\phi}|\tau_3 + |\eta|$, $\tilde{\Delta} = -|\vec{\phi}|\tau_3 + |\eta|$ are both matrices, and a Wick rotation was done to reach (.3b).

Converting to the Feynman integration variables, $\tilde{k} = k - \alpha p_1 + \beta p_2$, neglecting $p_1^2$, $p_2^2$ with respect to $\Delta^2$ and $\tilde{\Delta}^2$, performing the trace over the $\Gamma^\mu$ matrices and integrating over $\tilde{k}$-space, using

$$\int \frac{d^3\tilde{k}}{(2\pi)^3} \frac{1}{(k^2 + a)^3} = \frac{1}{32\pi a^{3/2}} \tag{.4}$$

we arrive at

$$j^\mu = \frac{1}{4\pi} \epsilon^{\mu\beta\gamma} p_\beta^1 p_\gamma^2 Tr\left[\int_0^1 dx \frac{x[|\eta| + |\phi|(2x - 1)\tau_3]}{(\sqrt{\phi^2 + \eta^2 + 2|\phi||\eta|(2x - 1)})^3}\right] \tag{.5a}$$

$$= \begin{cases} \dfrac{1}{4\pi} \epsilon^{\mu\beta\gamma} p_\beta^1 p_\gamma^2, & |\vec{\phi}| > |\eta|; \\ \dfrac{1}{8\pi} \epsilon^{\mu\beta\gamma} p_\beta^1 p_\gamma^2, & |\vec{\phi}| = |\eta|; \\ 0, & |\vec{\phi}| < |\eta|. \end{cases} \quad (.5b)$$

Where the trace in (.5a) is over the $\tau_3$ matrix, and (.5b) is derived by performing the integration in (.5a) for each individual element of the $\tau_3$ matrix and then adding to get the trace. The square root in (.5a) means taking the absolute value. Whenever the space-time dimension is odd one encounters a square root after $\vec{k}$-space integration.

From this result, $C = \dfrac{1}{8\pi} \Theta(|\vec{\phi}| - |\eta|)$ where

$$\Theta(x) = \begin{cases} 1, & x > 0; \\ \dfrac{1}{2}, & x = 0; \\ 0, & x < 0. \end{cases}$$

The discontinuity results from the limits taken: $|\vec{p}_1|, |\vec{p}_2| \ll |\vec{\phi}|, |\eta|$. Without taking the above limit, the asymptotic behavior of $j^\mu$ at $|\vec{\phi}| > |\eta|$ and $|\vec{\phi}| < |\eta|$ will still be the same but the transition will be continuous.

The induced fermion number of a baby skyrmion can be found by applying the result for $j^\mu$ on a baby skyrmion configuration:

$$\phi_3 = \cos(f(r))$$
$$\phi_2 = \sin(f(r))\sin\theta \quad (.6)$$
$$\phi_1 = \sin(f(r))\cos\theta$$

where

$$f(r = 0) = 0,$$
$$f(r = +\infty) = \pi \quad (.7)$$
$$r^2 = x^2 + y^2$$

The fermion number $Q$ is

$$Q_s = \int d^2x j_0$$

# Induced Quantum Numbers and Surface Currents

$$= \int_0^\infty 2\pi r dr \left[ 2C \frac{df}{dr} \frac{\sin f}{r} \right]$$

$$= - \int_{r=0}^{r=\infty} 4\pi C d(\cos(f(r)))$$

$$= \Theta(|\vec{\phi}| - |\eta|) \tag{.8}$$

When the $SU(2)$ and $U(1)$ symmetries are gauged, the induced current is determined as follows. First the partial derivatives in (.1) become the covariant derivatives in (3.3), then the second term in (3.3) is needed to conserve the current. The coefficient of these two terms are determined by the fact that when $g = 0$ it should reduce to (3.1). There is an additional term due to the coupling between the fermions and the $U(1)$ gauge field. It is calculated from the Reynman diagram (Fig. 2).

$$\langle \bar{\Psi} \Gamma^\mu \Psi \rangle = e^2 \int \frac{d^3 \vec{k}}{(2\pi)^3} Tr \left[ \Gamma^\mu \frac{1}{\not{k} - \not{p} - m} \Gamma^\beta \frac{1}{\not{k} - m} \right] \tag{.9a}$$

$$= e^2 \frac{1}{8\pi} \frac{\eta}{|\eta|} \Theta(|\eta| - |\vec{\phi}|) \epsilon^{\mu\beta\gamma} F_{\beta\gamma} \tag{.9b}$$

where in (.9a) $m = \eta \pm \phi_3$.

With the $SU(2)$ and $U(1)$ gauge fields, one can have stable vortices. (Strictly speaking, to have vortices one must break the symmetries completely. This requires additional Higgs fields. We have been assuming, implicitly, that these do not couple to the fermions.) A possible configuration of a vortex in $\vec{\phi}$ field is (.6), where

$$f(r = 0) = 0,$$
$$f(r = +\infty) = \frac{\pi}{2} \tag{.10}$$

and the gauge field has the asymptotic configuration

$$A_x^3 = \frac{y}{gr^2}$$

$$A_y^3 = -\frac{x}{gr^2} \tag{.11}$$

$$\vec{A}^{1,2} = 0$$

and is smooth at the origin. Applying (3.2) to this configuration, after similar calculation as above we find the induced fermion number for the vortex to be

$$Q_{v\phi} = \frac{1}{2} \Theta(|\vec{\phi}| - |\eta|) \qquad (.12)$$

Similarly, one finds

$$Q_{v\eta} = \frac{e\eta}{2|\eta|} \Theta(|\eta| - |\vec{\phi}|) \qquad (.13)$$

for the $U(1)$ vortices.

## ACKNOWLEDGMENTS

Y. H. Chen thanks Al Shapere for his help. F. Wilczek would like to thank the Smithsonian Institution for support as a Regents Fellow. We also thank the Wednesday movie club and E. Witten for helpful discussions. This work is supported in part by NSF grants No. Phy-82-17853 and Phy-87-14654.

## REFERENCES

1. M. Gell-Mann and M. Levy, *Nuovo Cimento* **16**, 705 (1960).
2. J. Goldstone and F. Wilczek, *Phys. Rev. Lett.* **47**, 986 (1981).
3. P. W. Anderson, *Science* **235**, 116 (1987).
4. I. Dzyaloshinskii, A. Polyakov, and P. Wiegmann, *Phys. Lett.* **127**, 112 (1988).
5. F. Wilczek and A. Zee, *Phys. Rev. Lett.* **51**, 2250 (1983).
6. Y. Aharonov and D. Bohm, *Phys. Rev.* **115**, 485 (1959).
7. V. Kalmeyer and R. B. Laughlin, *Phys. Rev. Lett* **59**, 2095 (1987); R. B. Laughlin, *Phys. Rev. Lett.* **60**, 2677 (1988).
8. A. Goldhaber, R. Mackenzie, and F. Wilczek, Harvard preprint, HUTP 88/044.
9. J. March-Russell and F. Wilczek, Harvard preprint, HUTP88/045.
10. A. N. Redlich, *Phys. Rev.* **D29**, 2366 (1984).
11. C. G. Callan and J. A. Harvey, *Nucl. Phys.* **B250**, 427 (1985).
12. J. Frohlich and P. A. Marchetti, "Quantum Field Theories of Vortices and Anyons," preprint 1988.
13. K. Wilson and J. Kogut, *Phys. Rev.* **C12**, 75 (1974).

C. H. Choi and Paul Muzikar
Department of Physics
Purdue University
West Lafayette, IN 47907

# The Superfluid Density Tensor in Unconventional Superconductors

We review the theory of the superfluid density tensor, $\rho_s$, for both singlet and triplet unconventional superconductors. This tensor is an important quantity characterizing the state of a superconducting metal; it is directly related to the magnetic penetration depth, and is a key factor in determining the energy of a vortex lattice. We include the effect of impurity scattering in our discussion, especially since such scattering has a pronounced effect on the anisotropy of $\rho_s$. We also allow for the effects of fermi surface anisotropy.

## I INTRODUCTION

The occurrence of unconventional order parameters in the heavy fermion or high $T_c$ superconductors is an exciting possibility. Theorists have devoted much effort recently to exploring the properties of unconventional superconductors;[1] this work is intrinsically interesting in its own right, and is of course crucial in helping experimentalists make an unambiguous identification of the types of order parameters existing in the various superconducting materials.

An unconventional order parameter $\Delta(\hat{k})$ is defined as one which has less rotational symmetry than the host lattice.[2] At least near $T_c$, $\Delta(\hat{k})$ will transform according to a particular nonidentity representation of the crystalline point group.

At lower temperatures, other representations may be mixed in. This definition of unconventionality thus relies on symmetry, and not on the pairing mechanism.

Often, it can be shown that order parameters transforming according to a particular irreducible representation must vanish on points or lines on the fermi surface.[3] In addition, an order parameter transforming according to a nonidentity representation must satisfy $\langle \Delta(\hat{k}) \rangle_{FS} = 0$; i.e. its angular average over the fermi surface vanishes. These two facts are of great importance.

In this paper we discuss the superfluid density tensor in unconventional superconductors, within the context of weak-coupling BCS theory. The superfluid density tensor $\rho_s$ is an important quantity for a variety of reasons. The magnetic penetration depth is directly related to $\rho_s$, and the energy of a vortex lattice depends on $\rho_s$.[4]

The superfluid density, $\rho_s$, will be a tensor quantity for two related reasons:

1. Anisotropy of normal state quantities such as the fermi velocity;
2. Anisotropy of the gap function $\Delta(\hat{k})$.

Various papers have reported calculations of $\rho_s$ or the penetration depth.[5,6,7,8,9] One interesting point, first noted by,[5] is that impurity scattering can significantly affect the anisotropy of $\rho_s$. Even for zero temperature and a spherical fermi surface, so that for a pure metal $\rho_s$ would be isotropic, the addition of an arbitrarily small concentration of impurities will make $\rho_s$ anisotropic when the gap has points or lines of nodes.

## II. IMPURITY SCATTERING IN UNCONVENTIONAL SUPERCONDUCTORS

In this section we briefly review the dramatic effects of impurity scattering when the order parameter is unconventional. This will serve to put in perspective our subsequent discussion of how the superfluid density is affected by such scattering.

Firstly, we note that for a gap transforming according to a nonidentity representation of the point group of the crystal, $T_c$ is substantially lowered by ordinary impurity scattering.[1,10] In fact, theory leads to results equivalent to the Abrikosov-Gorkov formula for the lowering of $T_c$ in conventional superconductors by paramagnetic impurities.

Second, in contrast to conventional superconductors, the density of states is strongly affected by ordinary impurity scattering.[1,10,11] For example, if the gap has a line of nodes on the fermi surface, the addition of an arbitrarily small concentration of impurities immediately makes the superconductor gapless; this means the density of states at zero energy is not zero.

Finally, superconducting properties can depend in complicated ways on $c$, the impurity concentration, and $v$ the scattering strength. This is in marked contrast to normal state properties, which often depend on $c$ and $v$ only through

$\tau$, the elastic time. For example, whether or not a state with point nodes is gapless depends not just on $\tau$, but on $c$ and $v$ separately.[12]

## III. QUASICLASSICAL THEORY

The quasiclassical version[13] of Gorkov's equations is perhaps the most efficient method for calculating many properties of superconductors, and is the approach we have used in computing $\rho_s$.[7,8] The central object of study is the quasiclassical propagator $\hat{g}(\epsilon, s, \vec{R})$; here, $\hat{g}$ is a $2 \times 2$ matrix in particle-hole space, $\epsilon$ is a Matsubara frequency, $s$ is a two dimensional variable indicating position on the fermi surface, and $\vec{R}$ indicates position in real space. Integrals over the fermi surface are denoted by $N(0) \int d^2s n(s)$, where $N(0)$ is the total density of states at the fermi surface, and $n(s)$ is normalized to unity: $\int d^2s n(s) = 1$.

In the presence of a spatially uniform superfluid velocity, we can remove the $\vec{R}$ dependence of $\hat{g}$ by performing a gauge transformation. The equations satisfied by the gauge-transformed propagator, $\bar{g}(\epsilon, s)$ are[14]

$$[(i\epsilon - Q)\hat{\tau}_3 - \bar{\Delta} - \bar{a}, \bar{g}] = 0, \qquad (1)$$

$$\bar{g} \cdot \bar{g} = -\pi^2. \qquad (2)$$

Here, $\bar{a}(\epsilon)$ is the impurity self-energy; it is given by $\bar{a} = c\bar{t}$, where $c$ is the concentration of impurities and $\bar{t}(\epsilon)$ is the impurity $T$ matrix. The $T$ matrix is computed in terms of $\bar{g}$ via the following equation

$$\bar{t} = v + N(0)v \int d^2s n(s) \bar{g} \bar{t}. \qquad (3)$$

For simplicity we have taken the impurity scattering to be purely $s$-wave of strength $v$; this is why $\bar{a}$ depends only on $\epsilon$, and not on $s$. The quantity $Q$ is given by $Q = m\vec{v}_F(s) \cdot \vec{v}_s$, where $\vec{v}_F(s)$ is the fermi velocity at point $s$ on the fermi surface and $\vec{v}_s$ is the superfluid velocity.

The off-diagonal pairing self-energy is given by $\bar{\Delta}(s)$, and is computed from $\bar{g}$ via the weak-coupling gap equation. It can be written as

$$\bar{\Delta} = i\Delta_1 \hat{\tau}_1 + i\Delta_2 \hat{\tau}_2. \qquad (4)$$

For spin singlet pairing $\Delta_1$ and $\Delta_2$ are real and even functions of $s$, while for triplet pairing they are real and odd functions of $s$. Note that this covers any separable, triplet gap. Since we are interested in computing $\rho_s$, we work to first order in $\vec{v}_s$; to this order, both the overall magnitude and the angular dependence of the gap are unaffected by the supercurrent. Effects higher order in $\vec{v}_s$ have been considered elsewhere.[14] The gap is strongly affected by the impurity scattering, and a self-consistent solution to the gap equation will take this into account.

The supercurrent is computed in terms of the propagator by

$$\vec{J} = 2N(0)T \sum_\epsilon \int d^2sn(s)\vec{v}_F(s)g_3, \qquad (5)$$

where $g_3$ is the $\hat{\tau}_3$ component of $\bar{g}$.

So we must solve self-consistently for $\bar{g}$, $\bar{\Delta}$, and $\bar{a}$ to first order in $\vec{v}_s$, and then use $g_3$ to evaluate the current. This gives us the superfluid density tensor. In general this involves some numerical work; close to $T_c$,[7,8] and close to $T = 0$,[5,9] more analytic progress is possible.

We now discuss the behavior of the impurity self-energy. For spin singlet pairing it is easy to check that $\bar{a}$ is an even function of $\vec{v}_s$. Hence, for the purpose at hand we may evaluate $\bar{a}$ at $\vec{v}_s = 0$. Furthermore, if the following two conditions hold for all real values of $x$:

$$\int d^2sn(s) \frac{\Delta_1(s)}{\sqrt{x^2 + \Delta_1^2(s) + \Delta_2^2(s)}} = 0, \qquad (6)$$

$$\int d^2sn(s) \frac{\Delta_2(s)}{\sqrt{x^2 + \Delta_1^2(s) + \Delta_2^2(s)}} = 0, \qquad (7)$$

then we have $a_1 = 0$ and $a_2 = 0$. Note that the above two conditions don't necessarily hold for an unconventional singlet gap, even if the following equations are true:

$$\langle \Delta_1 \rangle \equiv \int d^2sn(s)\Delta_1(s) = 0, \qquad (8)$$

$$\langle \Delta_2 \rangle \equiv \int d^2sn(s)\Delta_2(s) = 0. \qquad (9)$$

For triplet pairing the impurity self-energy has a somewhat different dependence on $\vec{v}_s$. It is straightforward to check that $a_3$ is an even function of $\vec{v}_s$, while $a_1$ and $a_2$ are odd functions of $\vec{v}_s$. Hence, we may write

$$a_1(\epsilon) = \vec{\gamma}_1(\epsilon) \cdot \vec{v}_s \qquad (10)$$

$$a_2(\epsilon) = \vec{\gamma}_2(\epsilon) \cdot \vec{v}_s. \qquad (11)$$

## IV. RESULTS AND DISCUSSION

For the spin singlet case we arrive at the following general formula for $\rho_s$:

$$\rho_{ij}^s = 2N(0)m\pi T \sum_\epsilon \int d^2s n(s) \frac{v_i^F v_j^F [(\Delta_1 - ia_1)^2 + (\Delta_2 - ia_2)^2]}{[(\epsilon + ia_3)^2 + (\Delta_1 - ia_1)^2 + (\Delta_2 - ia_2)^2]^{3/2}} \quad (12)$$

As discussed in the previous section, the $a_i(\epsilon)$ are evaluated at $\vec{v}_s = 0$, as are the $\Delta_i$. The $\Delta_i$ fully incorporate the effects of impurity scattering.

For the spin triplet case we get

$$\rho_{ij}^s = 2N(0)\pi T \sum_\epsilon \int d^2s n(s) \frac{m v_i^F v_j^F \Delta^2 + (\epsilon + ia_3)(\Delta_1 v_i^F \gamma_{1j} + \Delta_2 v_i^F \gamma_{2j})}{[(\epsilon + ia_3)^2 + \Delta^2]^{3/2}} \quad (13)$$

Here, we have defined $\Delta^2 = \Delta_1^2 + \Delta_2^2$. Recall that the $\vec{\gamma}_i(\epsilon)$ were defined in the previous section, and that $a_3(\epsilon)$ and the $\Delta_i$ are computed at $\vec{v}^s = 0$.

To evaluate these formulae at a general temperature will typically require numerical work. We refer the reader to recent publications for a look at graphs of some results.[7,8,9] These results make clear that the impurity scattering can have a large effect on the anisotropy of the superfluid density tensor. In contrast to many properties of a crystalline solid, which are made more isotropic by impurity scattering, $\rho_s$ can be made less isotropic.

In the Ginzburg-Landau limit, we can derive more explicit answers. We keep terms of order $\Delta^2$. First, consider the singlet case. We can solve the $T$ matrix equation to the required order and obtain the following result:

$$a_1 = \frac{i}{2\tau} \cdot \frac{\langle \Delta_1 \rangle}{|\epsilon|} \quad (14)$$

$$a_2 = \frac{i}{2\tau} \cdot \frac{\langle \Delta_2 \rangle}{|\epsilon|} \quad (15)$$

$$a_3 = \frac{i}{2\tau} sgn(\epsilon) \quad (16)$$

Here, $\tau$ is the normal state elastic scattering time defined by

$$\frac{1}{2\tau} \equiv \frac{N(0)\pi c v^2}{1 + (N(0)\pi v)^2}. \quad (17)$$

We then obtain the following formula for singlet gaps close to $T_c$

$$\rho^s_{ij} = 2N(0)\pi m T \sum_\epsilon \int d^2 s n(s) \frac{v^F_i v^F_j}{\left|\epsilon + \frac{sgn(\epsilon)}{2\tau}\right|^3} \left[\left(\Delta_1 + \frac{\langle\Delta_1\rangle}{2\tau|\epsilon|}\right)^2 + \left(\Delta_2 + \frac{\langle\Delta_2\rangle}{2\tau|\epsilon|}\right)^2\right]. \tag{18}$$

When $\langle\Delta_1\rangle = \langle\Delta_2\rangle = 0$ we then obtain the following simple expression

$$\rho^s_{ij} = 2N(0)m\pi T \sum_\epsilon \frac{1}{\left|\epsilon + \frac{sgn(\epsilon)}{2\tau}\right|^3} \int d^2 s n(s) v^F_i v^F_j [\Delta_1^2 + \Delta_2^2]. \tag{19}$$

This formula makes clear that for singlet pairing, when $\langle\Delta_1\rangle = \langle\Delta_2\rangle = 0$, the anisotropy of $\rho_s$ in the Ginzburg-Landau region is unaffected by impurity scattering; only the overall magnitude of the tensor is affected.

We now discuss the triplet case in the Ginzburg-Landau region. It is useful to define the following two vector quantities:

$$\vec{u}_1 = \int d^2 s n(s) \Delta_1(s) \vec{v}_F(s) \tag{20}$$

$$\vec{u}_2 = \int d^2 s n(s) \Delta_2(s) \vec{v}_F(s). \tag{21}$$

Solving for the $T$ matrix to the required order then gives

$$\vec{\gamma}_i = \frac{\vec{u}_i}{2\tau} \frac{m}{\left|\epsilon + \frac{sgn(\epsilon)}{2\tau}\right| \cdot |\epsilon|} \tag{22}$$

$$a_3 = -\frac{i}{2\tau} sgn(\epsilon) \tag{23}$$

We then arrive at the following formula for the superfluid density tensor

$$\rho^s_{ij} = A_{ij} + B_{ij} \tag{24}$$

with

$$A_{ij} = 2N(0)\pi mT \sum_\epsilon \frac{1}{\left|\epsilon + \frac{sgn(\epsilon)}{2\tau}\right|^3} \int d^2 sn(s) v_i^F v_j^F \Delta^2 \qquad (25)$$

and

$$B_{ij} = \frac{2N(0)\pi mT}{2\tau} \sum_\epsilon \frac{u_{1i}u_{1j} + u_{2i}u_{2j}}{|\epsilon|\left|\epsilon + \frac{sgn(\epsilon)}{2\tau}\right|^3}. \qquad (26)$$

So, in contrast to the singlet case, with triplet pairing the anisotropy will in general be dependent on the impurity scattering even in the Ginzburg-Landau limit.

For example, if we consider a polar phase gap, $\Delta(\vec{k}) = \Delta_0 \vec{k} \cdot \hat{z}$, with a spherical fermi surface, the superfluid density tensor will have two eigenvalues. In the limit of $T_c\tau \to 0$, the ratio of the two eigenvalues becomes

$$\frac{\rho_{zz}}{\rho_{xx}} \to -\frac{10}{3} \ln(4\pi\tau T_c). \qquad (27)$$

Thus the ratio can be made arbitrarily large by increasing the impurity concentration.

## V. ACKNOWLEDGMENTS

This work was supported in part by the Indiana Center for Innovative Superconductor Technology at Purdue University.

## REFERENCES

1. L. P. Gorkov, *Sov. Sci. Rev.* **A9**, 1 (1987).
2. D. Rainer, *Physica Scripta T* **23**, 106 (1988).
3. G. E. Volovik and L. P. Gorkov, *Sov. Phys.* JETP **61**, 843 (1985).
4. L. J. Campbell, M. M. Doria, and V. G. Kogan, *Phys. Rev. B* **38**, 2439 (1988).
5. F. Gross, B. S. Chandrasekhar, D. Einzel, K. Andres, P. J. Hirschfeld, H. R. Ott, J. Beuers, Z. Fisk, and J. L. Smith, *Z. Phys. B* **64**, 175 (1986).
6. A. J. Millis, *Phys. Rev. B* **35**, 151 (1987).

7. C. H. Choi and P. Muzikar, *Phys. Rev. B* **37**, 5947 (1988).
8. C. H. Choi and P. Muzikar, preprint.
9. R. A. Klemm, K. Scharnberg, D. Walker, and C. T. Rieck, *Z. Phys. B* **72**, 139 (1988).
10. K. Ueda and T. M. Rice, in *Theory of Heavy Fermions and Valence Fluctuations*, ed. by T. Kasuya and T. Saso (Springer, Berlin) 267 (1985).
11. L. P. Gorkov and P. A. Kalugin, *JETP Lett.* **41**, 253 (1985).
12. P. J. Hirschfeld, P. Wolfle, and D. Einzel, *Phys. Rev. B* **37**, 83 (1988).
13. J. A. X. Alexander, T. P. Orlando, D. Rainer, and P. M. Tedrow, *Phys. Rev. B* **31**, 5811 (1985).
14. C. H. Choi and P. Muzikar, *Phys. Rev. B* **36**, 54 (1987).

**C. Di Castro**
Dipartimento di Fisica
Università di Roma "La Sapienza"
I-00185 Roma, Italy

# Renormalized Fermi Liquid Theory for Disordered Electron Systems and the Metal-Insulator Transition

The basic concept of localization of electrons in a random potential was introduced by Anderson in 1958. Starting from the seventies a microscopic foundation to the theory of the underlying metal-insulator phase transition has been given via a renormalized perturbation theory.[1]

The electron-electron interaction modifies and completes[2] the Anderson picture, thus providing a richer theoretical framework[3] which, as it will be shown in the last section, is more suitable for the interpretation of the experiments.

The Anderson localization, being based on the one-electron picture, is specified once the amount of disorder present in the system is quantified by means of an effective coupling, which has been identified with the inverse conductance. The conductivity (or the diffusion constant) is renormalized and in strong disorder the system evolves towards an insulating phase.

When the mutual electron interaction is present several different scattering processes have to be considered. The related interaction amplitudes are strongly modified by the presence of disorder and additional effective couplings are required to specify the model.[4] The corresponding renormalization parameters have been identified in terms of physical quantities[5] appropriate to the Landau Fermi liquid theory, i.e. the Landau parameters, related to the interaction amplitudes, are renormalized by the disorder and acquire a scale dependence. By studying their renormalization group equations various universality classes of the metal-insulator transition driven by the electron mutual interaction in the presence of disorder are obtained.[3] The resulting theoretical picture summarized according to the result of Ref. 4 and 5, is analyzed in comparison with the recent experiments.

## I. ONE-ELECTRON PICTURE VERSUS EXPERIMENTS

In order to introduce some notation and to familiarize ourselves with the basic concepts of the theory we start by summarizing the one-electron picture.[1]

We start from the metallic phase and thus consider a free electron gas in the presence of a $\delta$-correlated random potential $U(x)$ defined via the averaging over the impurity distribution: $\overline{U(x)} = 0$, $\overline{U(x)U(x')} = \overline{u^2}\delta(x - x')$, $\overline{u^2}$ being proportional to the impurity concentration. In the Born approximation the scattering time $\tau_0$ is evaluated[6] to be $\tau_0 = \hbar(2\pi \overline{n^2} N_0)^{-1}$ where $N_0$ is the density of states for the free electron gas in $d$-dimensions, $N_0 \sim k_F^{d-2}/(\hbar^2/m)$.

Within the time $\tau_0$ between two collisions the electrons move along classical trajectories. No quantum interference effects are considered and the residual conductivity $\sigma_0$ has the Drude form

$$\sigma_0 = \frac{ne^2\tau_0}{m}, \tag{1.1}$$

where $n$ is the electron density.

The diffusion motion of the electrons implies that the density-density correlation function $\chi_{\rho\rho}$ has a diffusion pole in terms of the frequency $\Omega$ and the wave vector $q$

$$\chi_{\rho\rho} = -N_0 \frac{D_0 q^2}{-i\Omega + D_0 q^2} \tag{1.2}$$

with the diffusion constant $D_0 = (v_F^2 \tau_0/d)$.

In the static limit $\Omega = 0$, $q \to 0$ the density correlation function represents the change $\partial n/\partial \mu$ of the global particle density with respect to a static and homogeneous field, i.e. the chemical potential. In this simple case $\partial n/\partial \mu$ coincides with $N_0$. In the opposite limit ($q \to 0$, $\Omega \neq 0$), $\chi_{\rho\rho}$ vanishes, reflecting the conservation of the total number of particles. In the same way by the continuity equation one obtains the Einstein relation

$$\sigma_0 = \lim_{\Omega \to 0, q \to 0} e^2 \frac{\Omega}{q^2} \operatorname{Im} \chi_{\rho\rho} = e^2 N_0 D_0.$$

When we increase the disorder present, corrections to the Drude theory should appear. To evaluate these corrections in perturbation theory, we have to identify a suitable expansion parameter. Since we start from the weak disorder region a good dimensionless effective coupling is given by the ratio between $N_0\overline{u^2}$ (which specifies the amount of disorder present in the system) and the Fermi energy $\epsilon_F$ (which for the free electron system is the other dimensional quantity specifying the model). This ratio $N_0\overline{u^2}/\epsilon_F \sim \hbar/\epsilon_F\tau_0 \sim (k_F L)^{d-2}/g_0$ turns out to be inversely proportional to the conductance $g_0$ in units of $e^2/\hbar$.

$$g_0 = \frac{\sigma_0 L^{d-2}}{e^2/\hbar}, \qquad (1.3)$$

where $L$ is the characteristic size of the system. $t_0 \sim 1/g_0$ is thus the effective coupling of the model, as has been envisaged in the phenomenological theory. From now on we put $\hbar = e = 1$.

In the metallic phase the explicit dependence of $t_0$ on $L$ disappears in two dimensions, i.e. $t_0$ is marginal according to the critical phenomena terminology at $d = 2$ and acquires dimension $\epsilon = d - 2$ in terms of an inverse characteristic length at $d = 2 + \epsilon$.

As usual, marginality manifests itself as logarithmic singularities in perturbation theory. These logarithmic corrections to the Drude term at first order in $t_0$ physically correspond to the enhancement of the backscattering due to the quantum interference of the reverse patterns which are equivalent under time reversal.

The singularity appears as the logarithm* of the largest of the three ratios $\tau_L^{-1}/\tau_0^{-1}$, $\Omega/\tau_0^{-1}$ and $\tau_T^{-1}/\tau_0^{-1}$ where $\tau_L^{-1} \sim D_0 L^{-2}$ and $\tau_T$ is the inelastic scattering time $\tau_T \sim T^{-p}$ (the power $p$ depending on the choice of the possible scattering processes with $p = 2$ for the electron-electron interaction).

At $T = \Omega = 0$ we have

$$\sigma = \sigma_0(1 + t_0 \ln(D_0 L^{-2}\tau_0) + \cdots) \qquad (1.4)$$

giving rise to a decrease of $\sigma$ as $L$ increases. At $T = 0$ and for finite $\Omega$, $a \ln \Omega \tau_0$ appears in Eq. (1.4). At finite $T$, $a \ln T$ term in $\sigma$ in the weak disorder regime is also reproduced in many experiments on amorphous metallic films[7] (Fig. 1).

The logarithmic singularities, which at $d = 2$ dress the effective coupling $t_0$ also, as is standard in critical phenomena, are summed up to a power law behaviour in $d = 2 + \epsilon$. This is obtained by iterating the renormalization group equation up to the fixed point value $t^* \sim O(\epsilon)$ of the renormalized effective coupling $t$ via

---

*At $d = 3$ the perturbative correction term has a square root rather than logarithmic behaviour and in the weak disorder region the conductivity should assume the form $\sigma = \sigma_0 - mT^{p/2}$ with $m < $ - so that $\sigma$ decreases with decreasing temperature.

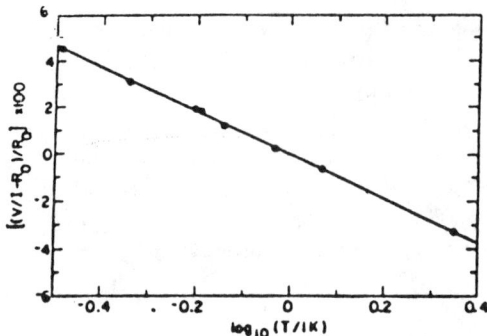

FIGURE 1 Resistivity rise plotted vs ln $T$ for PdAu film. Dolan and Osheroff. Ref. 7.

a continuous rescaling to infinity of the length ($L \to L' = sL$, $s \to \infty$). In $d = 2 + \epsilon$ the scaling theory of the Anderson transition is thus obtained. The behaviour of the system at the transition is governed by the divergence at $T = 0$ of the characteristic coherence length $\xi$ as a function of the disorder control parameter $\xi \sim (t_c - t_0)^{-\nu}$ where $\nu$ is the related critical index and $t_c$ specifies the critical value of the disorder. In the localized regime $\xi$ is the localization length. All the scaling quantities can be expressed in terms of the inverse of the coherence distance by means of their scaling dimension. The conductivity $\sigma$, for insxtance, scales as $\sigma \sim (t^*)^{-1} \xi^{-x_\sigma}$ with the scaling dimension $x_\sigma = d - 2 = \epsilon$ and therefore approaches zero at the transition with a power law behaviour in terms of the critical index $\mu$ satisfying the following scaling relation:

$$\sigma \sim (t^*)^{-1} \xi^{-x_\sigma} \sim (t_c - t_0)^\mu; \qquad x_\sigma = \frac{\mu}{\nu} = \epsilon. \qquad (1.5)$$

The density correlation function $\chi_{\rho\rho}$ has many singular correction terms. By implementing global and local conservation laws it can be shown that $\chi_{\rho\rho}$ has the same diffusive form (1.2), where, however, $D_0$ is replaced by the renormalized diffusion constant $D$, which scales as $\sigma$. According to scaling, the dimensions of $T$ and $\Omega$ must be the same as the dimension of $D\xi^{-2}$, i.e. $x_T = x_\Omega = 2 + \epsilon = d$. The critical index $\nu$, at least for the first few orders in the $\epsilon$-expansion, turns out to be $\nu = 1/\epsilon$ so that from Eq. (1.5) we have $\mu = 1$ for the general case of non-magnetic impurities.

Breaking the time reversal invariance or the spin conservation, the interference mechanism between time reversed patterns is changed. In the presence of magnetic impurities the interference appears at the order $t_0^2$ among complicated diffusive processes. A scaling theory of the transition is still obtained with $\nu = 1/2\epsilon$ and $\mu = 1/2$ instead of unity. If spin-orbit coupling is present an antilocalization term is obtained at the order $t_0$ and no phase transition appears in this context.

# Renormalized Fermi Liquid Theory for Disordered Electron Systems

In all the cases discussed above, thermodynamic quantities like the single particle density of states, the spin susceptibility and the specific heat do not show any critical behaviour.

A comparison with the experiments[8,9] shows that the theoretical investigation has to be extended to include the electron-electron interaction. We give here a very incomplete list of experimental results pointing in this direction:

1. Amorphous alloys like $Nb_x Si_{1-x}$[8] undergo a transition with a critical exponent $\mu = 1$, however, the single-particle density of states $N$ at the Fermi energy vanishes at the transition (Fig. 2), contrary to theoretical expectations.
2. The interference term leading to the index $\mu = 1$ in the Anderson localization is very sensitive to a time reversal invariance breaking term, due for example to an external magnetic field. In several systems as for instance in a-Si: Au[10] the measured value $\mu = 1$ is unchanged by the presence of the external magnetic field. The same result $\mu = 1$ is obtained when the transition is driven by a large external magnetic field,[11] which result cannot be accounted for by the previous theoretical predictions.
3. Uncompensated doped semiconductors like Si:P have a continuous transition[9] with an index $\mu = 0.51 \pm 0.05$. A possible explanation for this value of the index could be that the correlation between the electrons induces a quasi-static spin motion so that the spin-flip universality class with $\mu = 1/2$ should apply to this case. There are, however, many other experiments which imply the necessity of a full treatment of the interaction in the presence of disorder: a) The measured temperature-dependent term of the conductivity has a form

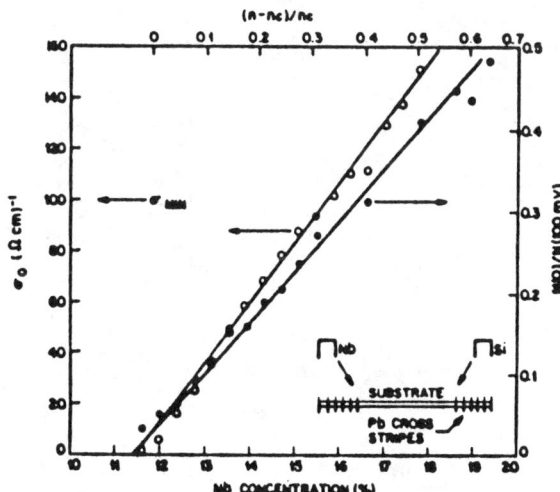

FIGURE 2 $\sigma$ and $N(0)/N(100\ mV)$ vs Nb concentration. Inset shows method of sample preparation. Ref. 8.

$$\sigma = \sigma_0 - mT^{1/2} \tag{1.6}$$

with a change of sign of $m$ with varying the donor concentration, which cannot be taken into account by the disorder correction term in $d = 3$. b) A strong temperature dependence of the spin susceptibility $\chi_s$,[12] Fig. 3a, and of the specific heat coefficient $\gamma = c_v/T$,[13] Fig. 3b, have been reported at low temperature.

4. When magnetic impurities are present a transition with an index $\mu = 1$ rather than $\mu = 1/2$ has been reported.[14] In the same way disordered Bi samples[15] and Si-doped $Al_x Ga_{1-x}$ As mixed crystals[16] show a metal-insulator transition with an index $\mu = 1$ in the presence of a strong spin orbit coupling like in a-Si:Au.[10] Although in the weakly disordered films the antilocalization effect of the one-electron picture has been observed,[15] the transition cannot be driven by pure disorder and another mechanism must occur.

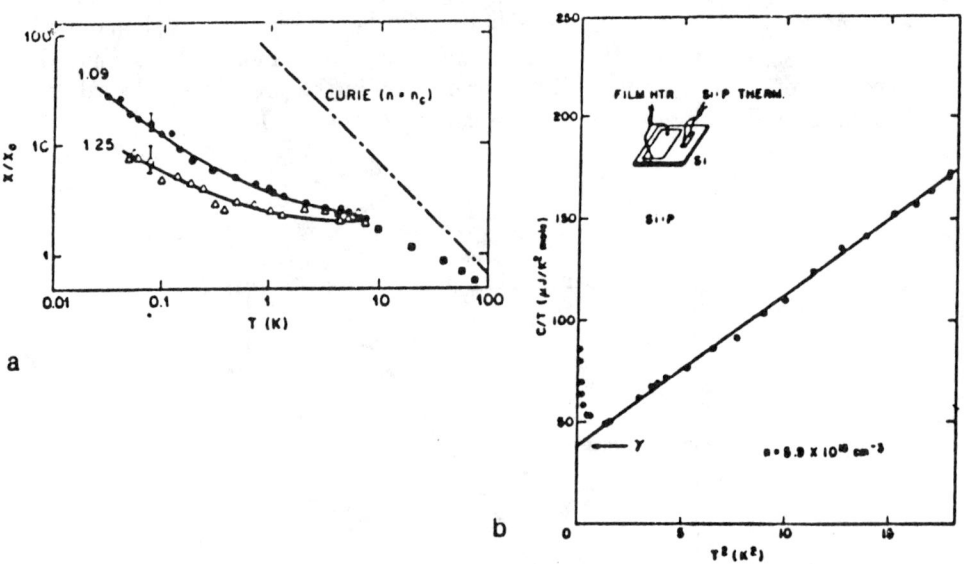

FIGURE 3a Spin susceptibility $\sigma_S$ normalized to the Pauli value. M. A. Paalanen et al. Ref. 12.

FIGURE 3b Specific-heat-to-temperature ratio, $C/T$, as a function of $T^2$, showing the low temperature deviation from the linear electronic specific heat. $\gamma = C/T$ assumes a $T$-dependence and deviates from the linear zero-$T$ intercept value indicated by the authors.

## II. EFFECTIVE FERMI LIQUID THEORY

In dealing with more realistic systems the electron-electron interaction has to be considered together with a random potential.

Approaching the problem from the metallic side, in the absence of disorder the effects of the interaction are taken into account by the Landau Fermi liquid theory. In the low temperature region and for excitation energy near the Fermi surface where the inelastic scattering time becomes extremely long, the quasi-particles are treated as if they were free but for a few effective parameters entering various equilibrium physical quantities like the thermodynamic density of states, the spin susceptibility and the specific heat

$$\frac{\partial n^L}{\partial \mu} = \frac{v^L/N_0}{1 + F_0^S} N_0, \qquad \chi_S^L = \frac{V^L/N_0}{1 + F_0^a} \chi_S^0, \qquad \frac{C_v^L}{C_v^0} = v/N_0, \qquad (2.1)$$

where $v^L/N_0 = m^*/m$, $m^*$ being the quasi-particle effective mass, and the superscripts $L$ and $0$ denote for each quantity the Landau form and the corresponding free-electron expression, respectively, i.e. $\chi^0 s = \mu_B^2 N_0$, $C_v^0 = \gamma_0 T = \pi^2/3 N_0 T$. The conventional Landau parameters $F_0^S$ and $F_0^a$ give a dimensionless measure of the interaction since they are related to the first coefficients of the expansion of the residual interaction among quasi-particles in terms of Legendre polynomials.

In the presence of disorder, repeated scattering processes with the impurities lead to a diffusive motion of the electrons and accordingly dress their mutual interaction into a frequency- and momentum-dependent form.

Singular correction terms, which are again logarithmic in $d = 2$, appear in several physical quantities.[2] A renormalized perturbation theory is then built. Together with the renormalized inverse conductance $t$ related to the disorder, the Landau parameters act as additional couplings, are renormalized and become scale dependent.

An effective Landau theory for disordered interacting electron systems is then obtained:[3,4,5]

$$\frac{\partial n}{\partial \mu} = Z_\rho \frac{\partial n^L}{\partial \mu}, \qquad \chi_S = Z_S \chi_S^L, \qquad C_v = Z C_v^L, \qquad (2.2)$$

where $Z_\rho$, $Z_S$ and $Z$ are clearly identified via Eq. (2.1) as the renormalization parameters of the interaction effective couplings and evolve according to a complicated set of renormalization group equations including $t$.

Historically this effective Landau picture has been obtained a posteriori. The interacting disordered electron system was mapped into a generalized non-linear σ-model in the first reference.[4] The renormalization parameters introduced to take care of the logarithmic singularities of this field theoretic model in $2 + \epsilon$ dimensions were afterward identified in terms of the modifications due to the combined effect of interaction and disorder on $\partial n/\partial \mu$, $\chi_s$ and $C_v$ leading to the modified Landau expressions (2.2). This was achieved[5] by implementing global and local conservation laws on the skeleton structure of the density, spin and energy correlation functions, which quite generally acquire a diffusive form

$$\chi_{\rho\rho} = -\frac{\partial n}{\partial \mu} \frac{D_0 q^2}{-i\Omega + D_\rho q^2}, \qquad \chi_{SS} = \frac{\chi_s D_s q^2}{-i\Omega + D_s q^2},$$

$$\chi_{EE} = -TC_v \frac{D_E q^2}{-i\Omega + D_E q^2}, \qquad (2.3)$$

with $\partial n/\partial \mu$, $\chi_S$ and $C_v$ given by Eqs. (2.2) and the particle, spin and heat transport coefficients renormalized as

$$D_\rho = \frac{D}{Z_\rho}(1 + F_0^s), \qquad D_S = \frac{D}{Z_S}(1 + F_0^a), \qquad D_E = \frac{D}{Z}, \qquad (2.4)$$

where $D$ is the diffusion constant appearing in the propagator of the diffusion mode of the theory.

Within this general framework the actual calculations have been carried out by considering that, as usual in the Landau scheme, the electron mutual interaction is effectively taken into account to all orders; the only expansion parameter of the theory is the coupling $t_0$ specifying the disorder.

The first perturbative correction term for the conductivity $\sigma$ in $d = 3$ gives a $T^{1/2}$ temperature dependence in the same form (1.6) as obtained from the experiments. This gives a first hint that the theory is going in the right direction at least in the weak disorder region. Moreover the coefficient $m$ depends on the various interaction effective couplings. Due to their evolution according to the group equations, the coefficient $m$ may acquire an additional temperature dependence even in the weak disorder regime,[17] as has been experimentally observed in GeSb,[18] or change sign according to the experimental results of Ref. 9.

The group equations[4,5] for the effective couplings have been obtained by evaluating the singular correction terms at first order in $t_0$, in $d = 2 + \epsilon$ dimensions and in the presence of a small time reversal symmetry breaking field to avoid additional complications due to the effect of the original quantum interference process on the electron mutual interaction.

The main consequences of the theory are now summarized and compared with the experiments.

The first general result of the perturbative analysis is that $\partial n/\partial \mu$ does not get renormalized and therefore $Z_\rho = 1$. $Z$ and $Z_S$ besides acting as effective couplings of the theory enter multiplicatively some of the relevant physical quantities as indicated in Eqs. (2.2) and (2.4). In spite of the fact that the thermodynamic density of states $\partial n/\partial \mu$ is not affected by the disorder in any relevant way, the density of states appearing in the renormalized specific heat coefficient

$\gamma$ is modified by $Z$ and behaves accordingly. The single particle density of states has a third different behaviour and in all the cases where a localization transition driven by the interaction will appear, it goes to zero in various complicated forms and a pseudo-gap appears.

The specific heat renormalization parameter $Z$ renormalizes also the temperature, whose scaling dimension is modified from the free electron value $x^0{}_T$,

$$x_T = x_T^0 - x_Z, \qquad (2.5)$$

where $x_Z$ is the scaling dimension of $Z$, and $x^0{}_T$, as we have seen assumes the value $d$ for the Anderson transition and the value 2 for the free metallic case.

Depending on the asymptotic values assumed by the couplings as the length rescaling parameter goes to infinity under different symmetry conditions, very different physical scenarios are obtained.

In the general case of non-magnetic impurities at $d = 2 + \epsilon$ and $T = 0$, there exists a critical line $t_0 = t_c(F_0^a, \epsilon)$ (which goes to zero in two dimensions where all the states should be localized) such that in the region of low disorder ($t_0 < t_c$) the system correctly scales to a conductor as the rescaling parameter goes to infinity, whereas in the region of strong disorder ($t_0 > t_c$), where the system should evolve towards a localized regime, $Z_S$ goes to infinity at a finite value of the rescaling parameter, signalling the presence of local magnetic moments. The group equations having been derived starting from the metallic regime cannot in this case be iterated to infinity.

At $t_0 = t_c(F_0^a, \epsilon)$ as the rescaling parameter goes to infinity, $Z$ and $Z_S$ diverge with a power law behaviour characterized by the scaling dimensions $x_Z = -3\epsilon$ and $z_{Z_s} = -4\epsilon$, respectively. At $t_0 = t_c$ and finite temperature, $\chi_S$ and $\gamma$ behave according to the following scaling forms:

$$\chi_S \sim T^{x_{Z_s}/x_T}, \qquad \gamma \sim T^{x_T/x_T}. \qquad (2.6)$$

Considering the values of $x_{Z_s}$ and $z_Z$ given above and the relation (2.5), $\chi_S$ and $\gamma$ diverge in this case as a function of the temperature

$$\chi_S \sim T^{-\frac{4\epsilon}{2+3\epsilon}} \underset{d=3}{\sim} T^{-4/5}, \qquad \gamma \sim T^{-\frac{3\epsilon}{2+3\epsilon}} \underset{d=3}{\sim} T^{-3/5}. \qquad (2.7)$$

We want to stress that $\chi_S$ and $\gamma$, being renormalized by two independent parameters, show different temperature behaviour.

In the group equation for the renormalized coupling $t$, the interaction introduces two competing terms. The divergence of $Z_S$ prevents the localizing one from being effective. $t$ approaches a fixed point value $t^* = 0$ with a scaling dimension $x_t = \epsilon$. Since $t \to 0$ the scaling low (1.5) is modified by the presence of the singular behaviour of $(t^*)^{-1}$ into $\mu = (\epsilon - x_t)\nu$. At the order considered here $\mu = 0$, i.e. $\sigma$ remains constant at the transition, whereas according to Eqs.

(2.4) the spin and energy diffusion constants go to zero as $Z$ and $Z_S$ go to infinity. Higher order correction terms could modify this result leading to a vanishing conductivity in agreement with the experiments[9] mentioned in the previous section.

Alternatively we can hypothesize that $t_0 = t_c (F_0^a, \epsilon)$ does not characterize the metal-insulator transition. The strong spin fluctuations, which are the relevant effect associated with $D_S \to 0$, lead instead to an instability line before the particle localization takes place. In this case the system, before reaching the instability, should make a cross-over to one of the universality classes with magnetic couplings which will be discussed later.

In any case the new relevant phenomenon characterized by the strong enhancement of $\chi_S$ and $\gamma$ is indeed present in the experiments[12,13] at temperatures below 1K, Figs. 3a and 3b. The measured values of $\chi_S$ and $\gamma$ also seem to suggest a stronger effect in $\chi_S$ than in $\gamma$.[19] When the temperature is sufficiently low that the spin scattering processes cut-off the effect of the Anderson term ($T < 100$ mK according to the first reference of 12) the agreement between theory (first formula (2.7)) and experiments appears to be more than qualitative.

Any mechanism which inhibits the spin fluctuations enhancement makes the localization term due to the interaction dominant in the equation for $t$. A finite fixed point value $t^*$ is reached and a bona fide phase transition driven by the interaction is obtained. Whenever a magnetic coupling, due to magnetic impurities, spin-orbit or strong magnetic field ($\mu_B H \gg kT$) is present, a finite $t^* \sim O(\epsilon)$ value is indeed obtained at first order in $t_0$. At the transition the ordinary scaling law $\mu$ - $\epsilon\nu$ is valid with the critical index for the conductivity $\mu = 1$.

In all the experiments[10,11,14,15,16] designed to meet the previous physical situations an index $\mu = 1$ is observed, thus solving the contradictions 2) and 4) of the list pointed out in the previous section while discussing the non-interacting theoretical framework in comparison with the experiments. With regard to point 2) the linear behaviour of $\sigma$ as a function of the disorder control parameter discussed here for the interacting case is not affected by the presence of a magnetic field, as is now confirmed by many experiments in systems with spin-orbit coupling.[10,15,16] See Fig. 4a. In the presence of spin-orbit coupling the localization term due to the interaction overcomes the antilocalization term due to the quantum interference, bringing back the localization transition with $\mu = 1$. Of course the approach to criticality and the critical concentration itself is modified by the presence of the external field as verified by the experiments,[15] Fig. 4a.

Although the critical behaviour of $\sigma$ is characterized by the same value of the index $\mu = 1$, the behaviour of $Z$ distinguishes among the previous cases three different universality classes of the metal-insulator transition driven by the interaction in the presence of disorder.

$Z$ approaches a finite fixed point value $Z^*$ in the presence of a strong magnetic field, i.e. its scaling dimension $x_Z$ vanishes in this case. The specific heat remains linear in temperature and the scaling behaviour of $\sigma$ as a function of temperature at criticality is $\sigma \sim T^{x_S/x_T} \sim T^{\epsilon/d}$.

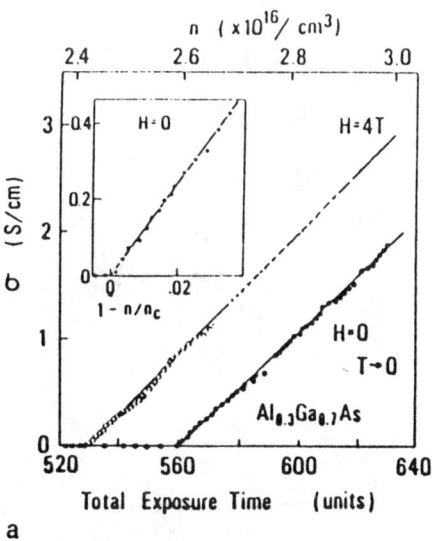

FIGURE 4a Extrapolated zero-temperature conductivity $\sigma(0, n)$ as a function of the total exposure time (or the electron concentration) in the magnetic field of 0 T and 4 T. $\sigma(0, n)$ near the metal-insulator critical point is plotted in the inset. Ref. 16.

As the scaling parameter goes to infinity, $Z$ approaches zero with a scaling dimension index $x_Z = \epsilon/2$ and $x_Z = \epsilon$ in the presence of magnetic impurities and spin-orbit coupling, respectively. Together with $D_\rho$, $D_E$ also goes to zero in the first case. In the spin-orbit case instead with $x_Z = x_\sigma = \epsilon$, $Z$ and $D$ vanish in the same way and $D_E$ according to Eq. (2.4) remains finite at the order considered.

$\gamma$ and $\sigma$ go to zero at criticality as a function of the temperature with a power law which is easily obtained by inserting the previous values of $x_Z$ in their scaling forms. Experiments on the specific heat are not yet available to check these predictions. In the spin-orbit case the behaviour of $\sigma$ at criticality as a function of the temperature is simply given by

$$\sigma \sim T^{x_\sigma/x_T} \sim T^{\epsilon/2} \underset{d=3}{\sim} T^{1/2}, \qquad (2.8)$$

where we have used the fact that in this case $x_T = d - x_Z = 2$. The $T^{1/2}$ behaviour is here a result of scaling at criticality and must not be confused with the analogous perturbative term valid in the weak disorder regime. The experiments on $Al_{0.3}Ga_{0.7}As$[16] show at criticality no deviation from a $T^{1/2}$ behaviour: see Fig. 4b.

We therefore conclude that whenever there is a magnetic coupling in the system, the predictions of the present theory for the disordered interacting electron systems fully agree with the experimental results. In the general case of

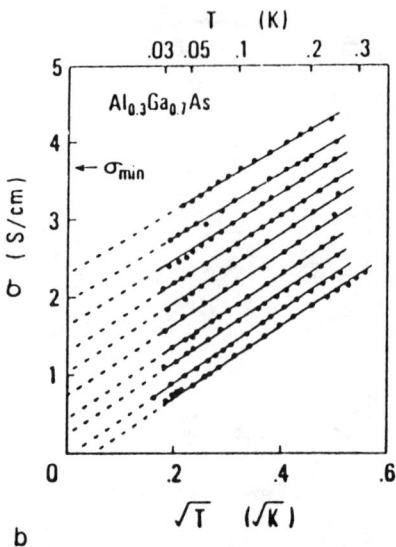

FIGURE 4b $\sqrt{T}$ dependence of the conductivity below 300 mK in the vicinity of the critical point. Solid lines are fits to the data and dotted ones indicate the extrapolation to 0 K. Ref. 16.

non-magnetic impurities, although the theoretical predictions of a strong enhancement of the spin susceptibility and of the specific heat coefficient $\gamma$ are indeed confirmed by the experiments, still more theoretical and experimental work is required to fully understand the transition and distinguish between the various hypotheses which have been suggested.

## REFERENCES

1. P. W. Anderson, *Phys. Rev.* **109**, 1492 (1958). For a review of Anderson localization see "Anderson Localization" Eds. Y. Nagaoka and H. Fukuyama, Springer-Verlag (1982) and Y. Nagaoka (ed.), *Progr. Theor. Phys. Suppl.* **84** (1985).
2. B. L. Altshuler and A. G. Aronov, *Solid State Comm.* **30**, 115 (1979); *Zh. Eksp. Teor. Fiz.* **77**, 2028 (1979); B. L. Atlshuler, A. G. Aronov, and P. A. Lee, *Phys. Rev. Lett.* **44**, 1288 (1980); H. Fukuyama, *J. Phys. Soc. Jpn.* **48**, 2169 (1980).
3. For a review see P. A. Lee and T. V. Ramakrishnan, *Rev. Mod. Phys.* **57**, 287 (1985). C. Castellani and C. Di Castro in "Applications of Field Theory to Statistical Mechanics" Ed. L. Garrido. Springer-Verlag (1985). C. Castellani, C. Di Castro and G. G. Strinati in "Fluctuations and Stochastic Phenomena in Condensed Matter." Ed. L. Garrido. Springer-Verlag (1986). Lo-

calization and Interaction. Ed. D. M. Finlayson, Scottish University Summer School in Physics (1986).
4. A. M. Finkel'stein, *Zh. Eksp. Teor. Fiz.* **84**, 168 (1983); *Sov. Phys. JETP* **57**, 97 (1983); C. Castellani, C. Di Castro, P. A. Lee and M. Ma, *Phys. Rev.* **B30**, 527 (1984); B. L. Altshuler and A. G. Aronov, *Solid State Comm.* **46**, 429 (1983).
5. A. M. Finkel'stein, *Pis'ma Zh. Eksp. Teor. Fiz.* **37**, 436 (1983) and *Z. Phys.* **B56**, 189 (1984); C. Castellani, C. Di Castro, P. A. Lee, M. Ma, S. Sorella and E. Tabet, *Phys. Rev.* **B30**, 1596 (1984); **B33**, 6169 (1986); C. Castellani and C. Di Castro, *Phys. Rev.* **B34**, 5935 (1986); C. Castellani, G. Kotliar and P. A. Lee, *Phys. Rev. Lett.* **59**, 923 (1987); C. Castellani, C. Di Castro, G. Kotliar, P. A. Lee and G. C. Strinati, *Phys. Rev. Lett.* **59**, 477 (1987) and *Phys. Rev. B* in press.
6. See for instance A. A. Abrikorov, L. P. Gor'kov and I. Ye Dzyaloshinskii, "Quantum Field Theoretical Method in Statistical Physics," Pergamon Press (1965).
7. G. J. Dolan and P. P. Osheroff, *Phys. Rev. Lett.* **43**, 721 (1979); S. Kobayashi, F. Komori, Y. Ootuka and W. Sasaki, *J. Phys. Soc. Jpn.* **49**, 1635 (1980).
8. G. Hertel, D. J. Bishop, E. G. Spencer, J. M. Rowell and R. C. Dynes, *Phys. Rev. Lett.* **50**, 743 (1983).
9. G. A. Thomas, M. Paalanen and T. F. Rosenbaum, *Phys. Rev.* **B27**, 3897 (1983); T. F. Rosembaum, R. F. Milligan, M. Paalanen, G. A. Thomas and R. N. Bhatt, *Phys. Rev.* **B27**, 7509 (1983).
10. N. Nishida et al., Solid State Electronics Proceedings of the Santa Cruz Conference (1984).
11. S. Von Molnar, A. Briggs, J. Flouquet and G. Remenyi, *Phys. Rev. Lett.* **51**, 706 (1983); R. M. Westervelt, M. J. Burns, P. F. Hopkins and A. J. Rimberg. This conference.
12. S. Ikeata and S. Kobayashi, *Solid State Comm.* **56**, 607 (1985); M. A. Paalanen, S. Sachdev and R. N. Bhatt, *Phys. Rev. Lett.* **57**, 2061 (1986); Y. Ootuka and N. Matsunaga, this conference.
13. G. A. Thomas, Y. Ootuka, S. Kobayashi and W. Sasaki, *Phys. Rev.* **B24**, 4886 (1981); S. Ikheata, private communication.
14. S. Okuma, F. Komori, Y. Ootuka, and S. Kobayashi, *J. Phys. Soc. Jpn.* **54**, 2382 (1985).
15. F. Komori, S. Okuma and S. Kobayashi, *J. Phys. Soc. Jpn.* **56**, 691 (1987) and this conference.
16. S. Katsumoto, F. Komori, N. Sano and S. Kobayashi, *J. Phys. Soc. Jpn.* **56**, 2259 (1987) and this conference.
17. C. Castellani, C. Di Castro, H. Fukuyama, P. A. Lee, and M. Ma, *Phys. Rev.* **B33**, 7277 (1986).
18. Y. Ootuko, S. Kobayashi, S. Ikehata, W. Sasaki and J. Kondo, *Solid State Comm.* **30**, 169 (1979).
19. M. A. Paalanen, private communication.

**Eduardo Fradkin**
Department of Physics
University of Illinois at Urbana-Champaign
1110 W. Green St., Urbana, IL 61801

# The Spectrum of Short-Range Resonating Valence Bond Theories

I review recent work done in collaboration with Steve Kivelson on the spectrum of short-range Resonating Valence Bond theories (RVB) of high temperature superconductors. I show that the Quantum Dimer Model (QDM), an approximation to short-range RVB, is equivalent to Polyakov's compact electrodynamics in 2 + 1 dimensions in the presence of sources. Using duality transformations and path-integral techniques I show that, in Euclidean space, this system is also equivalent to a three-dimensional Coulomb gas with phase factors which can be better interpreted as Berry phases. I argue that the ground state of the undoped system is unique, with a massive spectrum and, at the level of the QDM, permanent confinement of static holes. I also comment on the relevance of these results to superconductivity.

## I. INTRODUCTION

The discovery of the new superconductors[1] has spurred renewed interest in the behavior of strongly coupled electronic systems in two dimensions. The high critical temperatures, short correlation lengths (comparable with the lattice constant) and other features point toward a picture in which the interactions among electrons are indeed very strong.[2] In such situations it is doubtful that a Fermi liquid picture, with well defined quasiparticles ("dressed" electrons), would be of much use. Indeed there has been some speculation in recent literature[3] that Fermi liquid behavior may break down completely and that what one needs is an entirely different picture. However, it is very hard to determine when a system is actually a Fermi fluid. Many different definitions, none of them extremely precise, have been offered. Perhaps the clearest picture would be the existence of a sharp Fermi surface (FS). However, strong interactions can broaden the FS and overdamp the Fermi liquid quasiparticles. Thus it is quite possible that the Fermi liquid and the strong coupling limits may be continuously connected with each other without a phase transition separating both regimes.

In this lecture I adopt the strong coupling point of view. Here I report on work I have done recently in collaboration with Steve Kivelson (SUNY-Stony Brook) on the RVB picture of strongly correlated systems.[4] I will discuss mostly the behavior of the undoped system, which will turn out to be an insulator. I will not phrase my discussion in terms of your (or my) favorite variational wave functions. It has recently been shown, by extensive numerical studies,[5,6] that both RVB wave functions[7] and Neel states are quite competitive in terms of their energies (for a Heisenberg Model on a square lattice). The energy of a variational wavefunction is not a good indicator of how good the wavefunction is. The energy is indeed determined by the short distance properties. Wavefunctions with qualitatively different long distance behavior can have effectively very good variational energies if enough short range correlations are built in. What we need is to classify the generic properties of the possible phases of these systems in terms of the long distance, low energy, behavior of correlation functions which should be valid independently of your favorite choice of wavefunctions.

What we need to understand is the following:

(i) Do generic short-range theories with RVB-like[8,9] ground states exhibit a Mott-Superconductor transition.
(ii) What is their spectrum in the Mott phase?
(iii) How do we define a Mott phase?
(iv) What is their spectrum in the superconducting phase?
(v) Are there any massless (gapless) excitations in either phase?
(vi) We want to phrase all of this in terms of the long-range behavior of correlation functions.

## II. THE QUANTUM DIMER MODEL AND SHORT-RANGE RVB

I will begin by considering a very crude approximation to the strong coupling limit: the Quantum Dimer Model (QDM).[10] In this approximation, all species on the lattice are paired up in local nearest neighbor singlets (valence bonds). To a first approximation the non-orthogonality of the valence bond (VB) configurations will be neglected. These effects can be taken care of, for a short-range basis only,

# The Spectrum of Short-Range Resonating Valence Bond Theories

by orthogonalizing the basis. The result is a progressive non-locality of the Hamiltonian which will be discussed below. From now on the spin singlet states between neighboring sites (valence bonds) will be referred to as *dimers*. We will have the rule that a site on the square lattice will belong to *only one* VB. If there are no holes, all sites have exactly one VB touching them. In the presence of holes some sites will not belong to VB's.

I need to define an appropriate Hilbert space to label the states. Qualitatively speaking, we need to allow for fluctuations in VB occupancy of a given bond. Thus I define a Hilbert space on each link labeled by an integer, $\ell = 0, 1$ where $\ell = 0$ means no dimer and $\ell = 1$ indicates the presence of a dimer or VB on that link. Following Kivelson and Rokhsar[10] I define the Hamiltonian for the QDM to be

$$H = J \sum_{\text{plaquettes}} (|{=}{>}{<}|\,|\,|+|\,|\,|{>}{<}{=}|) \tag{1.1}$$

which represents tunneling processes (resonance) between VB's sharing the same plaquette of the lattice. I adopt this Hamiltonian as the definition of short-range RVB. Inclusion of orthogonalized states will make this Hamiltonian non-local, the main effect being resonance around loops $\Gamma$ on the lattice of arbitrary size $L$. The coupling constant is exponentially small

$$\left(\sim\left(\frac{1}{\sqrt{2}}\right)^L\right).$$

Their effect will be quite important.

## III. THE QUANTUM DIMER MODEL AND COMPACT ELECTRODYNAMICS

Let us now define a pair of operators $\tilde{A}_j$ and $\tilde{P}_j$ ($j = 1, 2$) at each link satisfying canonical equal-time commutation relations

$$[\tilde{A}_j(\tilde{x}), \tilde{P}_{j'}(\tilde{x}')] = i\delta_{jj'}\delta_{\tilde{x},\tilde{x}'} \tag{3.1}$$

where $(\tilde{x}j)$ labels the link $(\tilde{x}, \tilde{x} + \hat{e}_j)$. I require $\tilde{P}_j$ to have eigenvalues on the integers and hence $\tilde{A}_j$ has to be a phase ($0 \leq \tilde{A}_j < 2\pi$).

$$\tilde{P}_j(\tilde{x})|\ell_j(\tilde{x})\rangle = \ell_j(\tilde{x})|\ell_j(\tilde{x})\rangle \tag{3.2.a}$$

$$\langle \tilde{A}_j(\tilde{x})|\tilde{P}_j(\tilde{x})\rangle = \frac{1}{\sqrt{2\pi}} e^{i\ell_j(\tilde{x})A_j(\tilde{x})} \tag{3.2.b}$$

In order to recover the dimer Hilbert space I project onto $\ell_j$ by means of the link Hamiltonian

$$H_0 = \frac{1}{2\kappa} \sum_{\vec{r},j} \left[ \left( \tilde{P}_j(\vec{r}) - \frac{1}{2} \right)^2 - \frac{1}{4} \right] \tag{3.3}$$

In the limit $\kappa \to 0$ only the states with $\ell = 0, 1$ survive.

The effects of resonance can be described in terms of the full Hamiltonian

$$H = \frac{1}{2\kappa} \sum_{\text{links}} \left[ \left( \tilde{P}(\text{link}) - \frac{1}{2} \right)^2 - \frac{1}{4} \right]$$
$$+ 2J \sum_{\text{plaquettes}} \cos[\tilde{A}_j(\vec{R}) + \tilde{A}_j(\vec{R} + \hat{e}_2) - \tilde{A}_j(\vec{R}) - \tilde{A}_j(\vec{R} + \hat{e}_1)] \tag{3.4}$$

This Hamiltonian has to be supplied with the constraint

$$\tilde{P}_1(\vec{R}) + \tilde{P}_2(\vec{R}) + \tilde{P}_1(\vec{R} - \hat{e}_1) + \tilde{P}_2((\vec{R} - \hat{e}_2) = 1 - n(\vec{R}) \tag{3.5}$$

where $n(\vec{R}) = 1(0)$ if a hole is present (absent). These constraints commute (locally) with $H$. This is a gauge theory. The constraints are in fact the generators of local time-independent gauge transformations

$$|=> \to e^{i(\phi(1) + \phi(2) + \phi(3) + \phi(4)} |=> \tag{3.6}$$

where 1, 2, 3, 4 label the sites of the plaquette. We can put all this into a more transparent form by means of the following transformations. I first stagger the $\tilde{A}$ fields.

$$\tilde{A}_j(\vec{R}) = e^{i\vec{Q}\cdot\vec{R}} \tilde{A}_j(\vec{R}) \tag{3.7.a}$$

$$\tilde{P}_j(\vec{R}) = e^{i\vec{Q}\cdot\vec{R}} E_j(\vec{R}) \tag{3.7.b}$$

with $\vec{Q} = (\pi, \pi)$. In order to avoid double counting I demand

$$A_j(\vec{R}) = -A_{-j}(\vec{R} + \hat{e}_j) \tag{3.8.a}$$

$$E_j(\vec{R}) = -E_{-j}(\vec{R} + \hat{e}_j) \tag{3.8.b}$$

The constraint now takes the form

$$\Delta_j E_j = \Phi(\vec{R}) \tag{3.9}$$

where

$$\Delta_j f(\vec{R}) = f(\vec{R} + \hat{e}_j) - f(\vec{R}) \qquad (3.10.\text{a})$$

and

$$\Phi(\vec{R}) = e^{i\vec{Q}\cdot\vec{R}}[1 - n(\vec{R})] \qquad (3.10.\text{b})$$

In terms of these definitions I can write the Hamiltonian in the form

$$H = \frac{1}{2\kappa}\sum_{\vec{R},j}[E_j(\vec{R}) - \alpha_j(\vec{R})^2 - \alpha_j^2(\vec{R})] + 2J\sum_{\vec{R}}\cos\sum \tilde{A}_j(R) \qquad (3.11)$$

where $\sum \tilde{A}_j(R) = \Delta_1 A_2(\vec{R}) - \Delta_2 A_1(\vec{R})$ and $\alpha_j(R) = \frac{1}{2} e^{i\vec{Q}\cdot\vec{R}}$. The operator $\hat{Q}(\vec{R}) = \Delta_j E_j(\vec{R}) - \Phi(\vec{R})$ commutes with $H$ (locally) and is the generator of local time-independent gauge transformations. This is a gauge theory formally equivalent to compact electrodynamics.[11] There are four main differences: (1) We are not studying the vacuum sector but $\Delta \cdot E = \Phi$, (2) there is a background "electric" field $\alpha_j$, (3) it is anisotropic, and (4) the coupling constant $J$ has the wrong sign. We can take care of this last problem by shifting the $A$'s by $\pi$ on every other row of the lattice. This effectively shifts the $A$-flux by $\pi$ thus reversing the sign of $J$. While this procedure is harmless for the VB's, it does not make the holes very happy since it induces a frustration field. The gauge field $A_\mu$ should not be confused with the actual electromagnetic fields felt by the holes.

Polyakov[11] showed that the vacuum sector of this theory (in the absence of background "electric" fields) has a unique gauge invariant ground state with a purely massive spectrum, which can be interpreted as local fluctuations of flux. In particular, there is no "photon" (i.e. a non-singlet long wavelength excitation of the gauge fields). Static sources of the field were found to be permanently confined with an effective interaction potential which grows linearly with the separation between the sources. Polyakov found that the origin of this behavior, dramatically different from the predictions of perturbation theory, was the presence of a gas of magnetic monopoles (instantons) in the Euclidean path integrals. These monopoles were found to be always in a plasma phase which Polyakov argued was equivalent to confinement of static sources. In this lecture I will use the same techniques (see also ref. 12) to investigate our problem which involves a theory with sources and with background "electric" fields.

## IV. DUALITY TRANSFORMATION: QUANTUM FRUSTRATED DISCRETE GAUSSIAN MODEL

We can gain much insight on the behavior of the system by going to the dual system. This is done by solving the constraint (3.9) in terms of the integer—valued field $\hat{n}(\vec{R})$ residing on sites of the dual lattice. I will now show that the QDM is also equivalent as a quantum frustrated Discrete Gaussian Model (DGM) in 2 + 1 dimensions. I further show the equivalence between this model and a frustrated Sine-Gordon Model.

The constraint (3.9) is

$$\Delta_j E_j = \Phi$$

where $E_j$ has an integer-valued spectrum and $\Phi$ is an integer function. I can solve the constraint by defining at each link $(\vec{r}, j)$ of the direct lattice

$$\hat{E}_j(\vec{r}) = \epsilon_{jk}[\Delta_k \hat{n}(\vec{R}) + B_k(\vec{R})] \tag{4.1}$$

in terms of the operators $\hat{n}(\vec{R})$ and classical background fields $B_k(\vec{R})$ on sites and links of the dual lattice $\{\vec{R}\}$. The background field $B_k(\vec{R})$, which takes values only on the integers, has to satisfy

$$\epsilon_{jk}\Delta_j B_k(\vec{R}) = \Phi(\vec{r}) \tag{4.2}$$

This equation can be solved in some gauge. Note that the isotropic ("Lorentz") gauge is not available since the $B$'s *must* be integers. In fact we have the variance

$$n(\vec{R}) = n'(\vec{R}) + N(\vec{R}) \tag{4.3.a}$$

$$B_k(\vec{R}) = B'_k(\vec{R}) - \Delta_k N(\vec{R}) \tag{4.3.b}$$

where $N(\vec{R})$ is an integer-valued gauge transformation. I now define a set of operators $\hat{P}(\vec{R})$, on the dual lattice, which are canonically conjugated to $n(\vec{R})$.

$$[\hat{P}(\vec{R}), \hat{n}(\vec{R}')] = i\delta_{\vec{R},\vec{R}'} \tag{4.4}$$

Clearly, since $n(\vec{R})$ has its spectrum on the integers, $\hat{P}(\vec{R})$ must be a phase (i.e., $0 \leq \hat{P}(\vec{R}) < 2\pi$). We also have

$$e^{i\hat{P}}|n\rangle = |n+1\rangle \tag{4.5}$$

But the operator $e^{i\Sigma A_j}$, defined on a plaquette of the direct lattice, has the following in states $|\{\ell_j(\vec{r})\}>$

$$e^{\Sigma A_j}|\{\ell_j\}> = |\{\ell_j'\}> \quad (4.6)$$

where $\ell_j' = \ell_j + 1$ for links around the plaquette only. Thus both operators can be identified with each other.

$$e^{i\Sigma A_j}(\vec{r}) = e^{i\hat{P}(\vec{R})} \quad (4.7)$$

where $\vec{R}$ is the dual site to the right and above $\vec{r}$. By substituting (4.1) and (4.7) into (3.11), I find

$$H = \frac{1}{2\kappa} \sum_{\vec{r},j} \{[\Delta_j \hat{n}(\vec{R}) + B_k(\vec{R}) - \Gamma_k(\vec{R})]^2 - \Gamma_k^2(\vec{R})\} - 2J \sum_{\vec{R}} \cos \hat{P}(\vec{R}) \quad (4.8)$$

where

$$\Gamma_j(\vec{R}) = \epsilon_{jk}\alpha_k(\vec{r}) \equiv \frac{1}{2}\epsilon_{jk}e^{i\vec{Q}\cdot\vec{r}} \quad (4.9)$$

This is the Quantum Discrete Gaussian Model and it is frustrated due to the presence of the $B$'s. In fact the $\Gamma$'s carry zero curl. Quantum fluctuations originate from the last term in (4.8), the kinetic energy. This theory has a global symmetry $n(\vec{R}) \to n(\vec{R}) + m$, where $m$ is a constant integer. I solve (4.2) in the axial gauge, $B_1 = 0$. In this gauge I find

$$B_2(\vec{R}) = \frac{1}{2}(1 - (-1)^{r_1})(-1)^{r_2} \quad (4.10)$$

Of course other choices of gauge are equally acceptable. In this gauge the state $n(\vec{R}) = $ const. is equivalent to a column state, or spin Peierls state.[13,14] This can be seen most easily by considering the configurations of "electric" fields $E_j$. For $n(\vec{R}) = \bar{n}$ we have

$$E_j(\vec{r}) = \epsilon_{jk}B_k(\vec{R}) \equiv \epsilon_{jk}\frac{1}{2}(1 - (-1)^{r_1})(-1)^{r_2}\delta_{k,2} \quad (4.11)$$

which describes a column state. In fact, the different choices of gauge for solving (4.2) are in one-to-one correspondence with the classical dimer configurations.

The physics of (4.8) is very clear. The potential energy is infinite for any configuration which does not map into classical dimers covering the lattice without touching each other. The kinetic energy term simply induces tunneling processes among such degenerate states. The physical issue is whether these tunneling processes are important. Let us examine this problem from a path-integral point of view. This is done by going to imaginary time $\tau$ ($0 \leq \tau \leq \frac{1}{T}$, where $T$ is the temperature). Next I discretize the imaginary time axis with $N_\tau$ interval of length $\Delta t = \frac{1}{TN_t}$, and periodic boundary conditions (in time). The result is

$$Z = \sum_{\{n(\vec{r},\tau)\}} \exp\left\{-\frac{\Delta t}{2\kappa} \sum_{(\vec{r},\tau,j)}\right\}[(\Delta_j n(\vec{r},\tau) + B_j(\vec{r}) - \Gamma_j(\vec{r}))^2 - \Gamma_j^2(\vec{r})]$$

$$- \frac{1}{4J\Delta t} \sum_{\vec{r},\tau} (\Delta_\tau n(\vec{r},\tau))^2 \quad (4.12)$$

In (4.12) I have repeatedly used the result

$$<n|e^{2J\Delta t \cos \hat{P}}|n'> = \sum_{\ell=-\infty}^{+\infty} I_\ell(2J\Delta t)<n|e^{i\ell\hat{P}}|n'>$$

$$= I_{n-n'}(2J\Delta t) \approx \text{const } e^{-\frac{(n-n')^2}{4J\Delta t}}$$ (4.13)

where $I_\ell$ is a modified Bessel function.

I further use the result (valid for periodic boundary conditions)

$$\sum_{\vec{r},\tau,j} [(\Delta_j n + B_j + \Gamma_j)^2 + 2(\Delta_j n + B_j)\Gamma_j]$$

$$= \sum_{\vec{r}\tau j}' (\Delta_j n + B_j)^2 + 2 \sum [B_j \Gamma_j - n\Delta_j \Gamma_j] \quad (4.14)$$

We can now make use of the fact that

$$\Gamma_j = \epsilon_{jk}\alpha_j = \frac{1}{2}\epsilon_{jk}(-1)^{x_1+x_2} \equiv \frac{1}{2}\epsilon_{jk}\Delta_k\Lambda(\vec{r}) \quad (4.15)$$

where $\Lambda = 0$ for $x_1 + x_2$ even (odd). Thus we have

$$\Delta_j \Gamma_j = 0 \quad (4.16.a)$$

$$\sum_{\vec{r}} B_j \Gamma_j = \sum_{\vec{r}}' \frac{1}{2}\epsilon_{jk}(\Delta_k\Lambda)B_j \equiv -\sum_{\vec{r}}' \frac{1}{2}\epsilon_{jk}\Delta_k B_j \Lambda = \frac{N}{2} N_t \quad (4.16.b)$$

This last result is valid in the absence of holes ($N = N_{\text{sites}}$). Thus (4.14) becomes

$$\sum_{\vec{r},\tau,j} [(\Delta_j n + B_j + \Gamma_j)^2 - \Gamma_j^2] = \sum_{j,\vec{r},\tau} (\Delta_j n + B_j)^2 + NN_t \quad (4.17)$$

We can now write

$$Z = \sum_{\{n(\vec{r},\tau)\}}{}' \exp - \left[\frac{\Delta t}{2\kappa}\sum_{\vec{r},\tau,j}(\Delta_j n + B_j)^2 - \frac{N}{2\kappa T} - \frac{1}{4J\Delta t}\sum_{\vec{r},\tau}(\Delta_\tau n)^2\right] \quad (4.18)$$

with PBC's. Please keep in mind that one should only consider the limit $\kappa \to 0$.

The partition function of Eq. (4.18) represents a classical frustrated discrete Gaussian model on a three-dimensional Euclidean lattice, as expected.[15] Polyakov[11] and, independently, Kosterlitz[16] showed that the unfrustrated version of this system exists only in one phase: the "smooth" phase. In this phase the correlation function $C_\alpha(\vec{R})$

$$C_\alpha(\vec{R}) = \langle \exp[i\alpha(n(\vec{R}) - n(\vec{O}))]\rangle_{\mathrm{DGM}} \quad (4.19)$$

has long-range order, i.e. as $|\vec{R}| \to \infty$ (for $\kappa, j$ finite)

$$C_\alpha(\vec{R}) \approx \mathrm{const} + e^{-|\vec{R}|/\xi} \quad (4.20)$$

This state has $n(\vec{R}) \approx n_0$, same constant value. If this were the case for the *frustrated* model, it would imply, in the gauge $B_1 = 0$, a *column phase* for the dimers. The classical unfrustrated model has, up to a symmetry operation (shift by an arbitrary integer), a unique ground state. The frustrated model has an infinite degeneracy of the classical ground state. The degeneracy grows like $2^{N_s}$, where $N_s$ is the number of sites in the space direction. It is, however, independent of the size in the imaginary time direction. This means that if the boundary conditions are fixed in some way, tunneling histories will boast *at most* an action proportional to the number of tunneling events (i.e. flips from one ground state to the other) in that history with amplitude proportional to $\exp\left(-\frac{1}{4J\Delta t}\right)$ per event. There still is a very large entropy for such events ($\ell n (N_s N_t)$). For histories with a finite fraction of their lifetime with tunneling events present, the action is $\exp\left[\frac{-fN_t}{4J\Delta t}\right]$ where $f$ is the fraction. For $N_t\Delta t = \frac{1}{T}$ very large (i.e. very low temperatures), these factors become very small and always beat the entropy. For *any* non-zero temperature entropy wins and tunneling events proliferate thus disordering the ground state. We conclude that at $T = 0$ some sort of long-range order is present but it is unstable against thermal fluctuations. Something very similar is known to happen in other classical frustrated three-dimensional models, such as the Ising antiferromagnet on an fcc lattice.

We can gain further insight by considering the mapping to the Sine-Gordon theory. This is done by softening the configuration from integers $\{n\}$ to a continuous variable $\{\phi\}$ with a periodic cosine potential

$$Z_{DGM} \approx Z_{SG} = \int D\phi \, e^{-S_{SG}[\phi]} \tag{4.21}$$

$$S_{SG}[\phi] = \frac{1}{2J\Delta t} \sum_{(\vec{r},\tau)} (\Delta_\tau \phi)^2 + \frac{\Delta t}{2\kappa} \sum_{\vec{r},\tau,j} (\Delta_j \phi + B_j)^2 + y \sum_{\vec{r},\tau} \cos(2\pi\phi(\vec{r},t)) \tag{4.22}$$

This Sine-Gordon theory is also frustrated and thus the equal perturbative treatment in powers of $y$ must be considered with great care for it leaves out the tunneling events described above. These configurations connect the different classically solutions (i.e. saddle points).

## V. MONOPOLES, TUNNELING AND BERRY PHASES

We can also describe this problem in terms of topological excitations. These *instantons* are Polyakov's monopoles. Such a description can be easily achieved by means of a Poisson Summation Formula (PSF). Consider Eq. (4.18) and apply the PSF to it.

$$\sum_{n=-\infty}^{+\infty} f(n) = \int_{-\infty}^{+\infty} d\phi \sum_{m=-\infty}^{+\infty} f(\phi) e^{i2\pi m\phi} \tag{5.1}$$

The result is

$$Z = \int D\phi \sum_{\{m\}} \exp\left[ -\frac{1}{4J\Delta t} \sum (\Delta_\tau \phi)^2 - \frac{\Delta t}{2\kappa} \sum (\Delta_j \phi + B_j)^2 \right.$$
$$\left. - \frac{N}{2\kappa T} + i \sum 2\pi m \phi \right] \tag{5.2}$$

where $\phi = \phi(\vec{r}, \tau)$. By integrating out the $\phi$-fields one finds

$$Z = \tilde{Z}_{CG} e^{-\frac{N}{2\kappa T}} \exp\left[ -\frac{1}{2} \sum_{\vec{x},\vec{x}'} \epsilon_{\alpha\mu\lambda} \Delta_\lambda^x B_\mu(x) G_0(\vec{x} - \vec{x}') \epsilon_{\alpha\nu\rho} \Delta_\nu^{x'} B_\rho(x') \right] \tag{5.3}$$

where $G_0(\vec{x}, \vec{x}')$ is the lattice Green function for the three-dimensional anisotropic problem

$$\left[ \frac{1}{2J\Delta t} \Delta_\tau^2 + \frac{\Delta t}{\kappa} \sum_{j=1,2} \Delta_j^2 \right] G_0(\vec{x}, \vec{x}') = \delta_{\vec{x},\vec{x}'} \tag{5.4}$$

and $\tilde{Z}_{CG}$ is the position function for the Coulomb (or monopole) gas

# The Spectrum of Short-Range Resonating Valence Bond Theories

$$\tilde{Z}_{CG} = \sum_{\{m(\vec{r},\tau)\}} \exp\{2\pi^2 \sum_{x,x'} m(x)G_0(\tilde{x}, \tilde{x}')m(x')i \sum_x 2\pi m(x)\varphi(x)\} \quad (5.5)$$

The phase $\varphi(x)$ is given by

$$\varphi(x) = \sum_{x'} G_0(\tilde{x}, \tilde{x}')\Delta_\mu^{x'} B_\mu(x') \quad (5.6)$$

The solution to Eq. (5.4) is ($N, N_t \to \infty$)

$$G_0(x, x') = \int \frac{d^3q}{(2\pi)^3} \frac{\left(-\frac{1}{4}\right)e^{iq(x-x')}}{\frac{1}{2J\Delta t}\sin^2\left(\frac{q^0}{2}\right) + \sum_{j=1,2}\frac{\Delta t}{\kappa}\sin^2\left(\frac{q_j}{2}\right)} \quad (5.7)$$

At long distances (and long times) $G(x, x')$ becomes, up to an anisotropy, equivalent to the three-dimensional Coulomb interaction, i.e. $G(R) \sim \frac{1}{R}$. The phase factor in (5.5) is the Berry phase of the ground state. It is determined by the distribution of $\Delta_\mu B_\mu(x)$. At first sight this is surprising since it appears to be gauge-dependent. However, under an *integer-valued* gauge transformation $B_\mu \to B_\mu + \Delta_\mu N(r)$ we get

$$\varphi(x) \to \varphi(x) + \sum_{x'} G_0(x, x')\Delta_x^2 N(x')$$
$$= \varphi(x) - \sum_{x'} \Delta_x^2 G_0(x, x')N(x') = \varphi(x) - N(x) \quad (5.8)$$

Since $\varphi$ enters in the form $e^{i\pi m\varphi}$ a shift of $\varphi$ by any integer is allowed. Thus the phase $\varphi(x)$ is defined up to an integer-valued shift and, in this sense, gauge-invariant. Note that *it is not* invariant under non-integer gauge transformations. This is a consequence of the symmetries. The phases $\varphi(x)$ take different values on the four distinct sublattices. An explicit calculation yields a phase $\theta(x) \equiv 2\pi\varphi(x)$ to be

$$\theta(x) = \begin{cases} -\frac{\pi}{4} & x_1 \text{ even, } x_2 \text{ even} \\ +\frac{3\pi}{4} & x_1 \text{ odd, } x_2 \text{ odd} \\ +\frac{\pi}{4} & x_1 \text{ odd, } x_2 \text{ even} \\ -\frac{3\pi}{4} & x_1 \text{ even, } x_2 \text{ odd.} \end{cases} \quad (5.9)$$

Thus the Quantum Dimer Model (undoped), and hence the Quantum Frustrated Discrete Model, turns out to be equivalent to a 3D Coulomb Gas with phase factors (5.9) ("Berry phases").

Little is known about the behavior of such Coulomb gases. The phase factors are reminiscent of topological terms. However, they *do not* couple to the total topological charge on the total system but only on a particular sublattice. In the absence of the phase factors, we know, after Polyakov's work, that the Coulomb gas is always in a plasma phase. The phase factors suppress configuration with wild global fluctuations of the total charge on each sublattice. In three dimensions this does not appear to be a very stringent requirement for a Coulomb gas which behaves like a plasma. Thus we do expect most of Polyakov's analysis to remain valid. This means confinement of static sources (holes). The sublattice structure also means that dimer-dimer correlation functions, which can be related to dipoles in the Coulomb gas language, will exhibit long-range order. This is discussed in detail elsewhere.[17] We also have numerical evidence in support of this picture.

## IV. CONCLUSIONS AND OUTLOOK

In these lectures I presented a picture of undoped short-range RVB systems in terms of the properties of an Abelian Gauge Theory with sources and its dual theories: the discrete frustrated Quantum Gaussian Model and 3D Coulomb gases with Berry phases. I left out of this picture several important ingredients which are vital for a connection with spin systems and superconductors: (a) non-orthogonality of the VB states, (b) holes and (c) spinons. In another set of lectures in this volume, Kivelson argues that non-orthogonality can be put back in by adding larger loops in which the VB's resonate. I argue that such effects can effectively destroy the long-range order. Qualitatively such effects are similar to adding an extra matter field (the "orthogonalizer") which will induce the large loops. In the presence of such dynamical fields the string between two static holes "will break" at some distance by creation of a pair of such excitations. Spurious and dynamical holes do very much the same thing. The long-range orientational order of the VB's is a little harder to destroy. However, any thermodynamic density of holes can effectively do it. I also argued above that thermal fluctuations (tunneling events) can also destroy this state. Thus I'd like to argue that although long-range order of the VB's appears to be natural in our scheme, it can be ruined quite effectively by excitations such as holes or spinons. Thus I conclude that short-range RVB theories have the following generic behavior:

(a) the ground state is unique
(b) the spectrum has a finite gap
(c) all states are singlets
(d) static holes are linearly confined at the level of the Quantum Dimer Model but

(e) dynamical holes can be free and massive and appear as gauge singlets
(f) dynamical holes pick up a "frustration" energy as well as a Bohm-Aharonov phase factor due to their coupling to the gauge fields
(g) this last effect may effectively bind the holes in red-black pairs
(h) we also have evidence of a Meissner effect for electromagnetic gauge fields (i.e. superconductivity)

Note: Some of my conclusions for the undoped system were also found by N. Read and S. Sachdev in the context large-N antiferromagnets. See Read's lecture elsewhere in this volume.

## ACKNOWLEDGEMENTS

I'd like to thank Steve Kivelson, my collaborator in this work, for sharing his insights on this subject with me. This work began during our participation at the Aspen Workshop on Strongly Correlated Systems. I'd like to thank the Aspen Center for Physics for its kind hospitality. I also want to thank Professor Zlatko Tesanovic for organizing this wonderful workshop at Johns Hopkins. This work was supported in part by NSF Grant DMR84-15063.

## REFERENCES

1. J. G. Bednorz and A. Müller, *Z. Phys. B* **64**, 188 (1987).
2. For a review of experimental properties, see the recent review by D. M. Ginsberg, "Physical Properties of High Temperature Superconductors I," World Scientific, Singapore (1989).
3. P. W. Anderson, Princeton Preprint (1989).
4. Much of the contents of this lecture can be found in a series of papers I am preparing in collaboration with S. Kivelson. We also discuss the behavior of holes and superconductivity.
5. S. Liang, B. Doucot and P. W. Anderson, *Phys. Rev. Lett.* **61**, 365 (1988).
6. D. Huse and V. Elser, *Phys. Rev. Lett.* **60**, 2531 (1988).
7. P. W. Anderson, *Science* **235**, 1196 (1987).
8. S. Kivelson, D. Rokhsar and J. Sethna, *Phys. Rev. B* (R.C.) **35**, 8865 (1987).
9. B. Sutherland, *Phys. Rev. B* **37**, 3786 (1988).
10. D. Rokhsar and S. Kivelson, *Phys. Rev. Lett.* **61**, 2376 (1988).
11. A. M. Polyakov, *Nucl. Phys. B* **120**, 429 (1977).
12. T. Banks, R. Myerson and J. Kogut, *Nucl. Phys. B* **129**, 493 (1977).

13. I. Affleck and J. B. Marston, *Phys. Rev. B* **37**, 3774 (1988).
14. T. Dombre and G. Kotliar, MIT Preprint (1988).
15. D. Fradkin and L. Susskind, *Phys. Rev. D* **17**, 2637 (1978).
16. J. M. Kosterlitz, *J. Phys. C* **10**, 3753 (1977).
17. E. Fradkin and S. Kivelson, in preparation, and E. Fradkin, S. Kivelson and S. Liang, in preparation.

B. A. Jones
Lyman Laboratory of Physics
Harvard University
Cambridge, MA 02138

# Antiferromagnetic Phase Instability in the Two-Impurity Kondo Problem

We examine a model of two spin-one-half local moments in a conduction electron sea using Wilson's methods of numerical renormalization group. The competition between correlations of the itinerant electrons with the local moments, and of the local moments with each other, gives rise to complex behavior. For antiferromagnetic inter-moment interactions there occurs an unstable fixed point at finite coupling. This fixed point cannot be expressed in terms of a free-electron based Fermi liquid, a conclusion based upon examination of the fixed-point energy levels and of other anomalous properties of the system near the critical point.

## I. INTRODUCTION

The competition between Kondo effects and magnetic ordering in a lattice can give rise to a complex set of interactions, effects which will have bearing on the behavior of heavy fermion materials and also perhaps on that of some high-temperature superconductors. The full theoretical solution of lattice of localized moments is at this point still far from complete, however.

As a start towards understanding this problem, we have been analyzing the behavior of two spin-$\frac{1}{2}$ local moments ("impurities") in a sea of conduction electrons, using Wilson's methods[1,2] of numerical renormalization group. A review of previous studies[3] of the two-impurity system appears in Ref. 4. The two pertinent energy scales are the single-impurity Kondo temperature $T_K$ and the inter-impurity RKKY coupling $I_o$. $T_K$ is that temperature scale that would govern the quenching of the local moment spins in the absence of any interactions between them, $I_o = 0$. The RKKY interaction is generated indirectly by spin-flip scattering of a conduction electron off of one impurity site to the other, and can be of either sign depending on inter-impurity separation.

The behavior of the system for ferromagnetic RKKY interactions between the local moments is described in Refs. 5 and 6. In this paper we wish to focus on the case of antiferromagnetic interactions, and in particular on an unstable fixed point which occurs at finite coupling, between two stable regimes. At the unstable point the staggered susceptibility and the linear coefficient of specific heat diverge, and a simple explanation in terms of Fermi liquid theory breaks down.

We will describe how the preceding analysis is based upon examination of the fixed-point levels, and in the process illustrate the complexity of the unstable state. The paper is organized as follows. The second section describes the transformations of the original Hamiltonian into an iterative form appropriate for the numerical renormalization group procedure. The Hamiltonian is first linearized, and then expressed in terms of operators on a logarithmically discretized space. The third section describes the stable fixed points for the antiferromagnetic regime, and their classification in terms of one-particle Hamiltonians of a set of $N$ free electrons. We then present the lowest energy levels of the unstable fixed point, and discuss the contradictions inherent in trying to reproduce them by filling a set of single-particle levels. In the fourth section we briefly describe how finite-iteration deviations of the energy levels from their fixed-point values can be quantitatively parameterized by a Fermi-liquid-like effective Hamiltonian. The variation of the "Fermi-liquid" parameters as a function of initial energy scales, as well as the thermodynamic quantities which can be calculated from the effective Hamiltonian, give additional characterization of the unstable state. We end with a summary and directions of future research.

## II. DISCRETIZATION OF THE HAMILTONIAN

The Hamiltonian is an extension of the original Kondo model to two spin-$\frac{1}{2}$ impurities $\vec{S}_1$ and $\vec{S}_2$, separated by a distance $r$. Direct interactions, of strength $J$, are only with those conduction electrons $\left(\text{also of spin }\frac{1}{2}\right)$ at each site:

$$H = \int d^3k \epsilon_k a_{k\mu}^\dagger a_{k\mu} + J[\vec{s}_c(\mathbf{r}_1)\cdot\vec{S}_1 + \vec{s}_c(\mathbf{r}_2)\cdot\vec{S}_2] \tag{1}$$

where

$$\vec{s}_c(\mathbf{r}_i) = \frac{\Omega_o}{8\pi^3} \int d^3k \int d^3k' e^{i(\mathbf{k}'-\mathbf{k})\cdot \mathbf{r}_i} a^\dagger_{\mathbf{k}\mu} \frac{1}{2} \vec{\sigma}_{\mu\mu'} a_{\mathbf{k}'\mu'}.$$

Here $\Omega_o$ is the volume and $J > 0$.

To apply the Wilson procedure, we assign to the conduction electrons a linear dispersion, a constant density of states $\rho$, and a band from $-D$ to $D$. The Hamiltonian (1) can then be linearized and all conduction electron states assigned a parity even (e) or odd (o) with respect to exchange of the two sites. As a last approximation, we take the resulting energy-dependent coupling coefficients to be constants just depending on the impurity separation, which enforces particle-hole symmetry on the interactions. (Note that this is not equivalent to evaluating all the coupling coefficients at a specific energy such as the Fermi energy, an approximation which would generate only ferromagnetic RKKY interactions.) The resulting Hamiltonian is

$$\frac{H}{D} = \int_{-1}^{1} d\epsilon\, \epsilon [a^\dagger_{\epsilon e\mu} a_{\epsilon e\mu} + a^\dagger_{\epsilon o\mu} a_{\epsilon o\mu}] + \frac{H_{int}}{D}, \quad (2)$$

$$\frac{H_{int}}{D} = \rho J_e \int_{-1}^{1} d\epsilon \int_{-1}^{1} d\epsilon' a^\dagger_{\epsilon e\mu} \frac{1}{2} \vec{\sigma}_{\mu\mu'} a_{\epsilon' e\mu'} \cdot (\vec{S}_1 + \vec{S}_2)$$

$$+ \rho J_o \int_{-1}^{1} d\epsilon \int_{-1}^{1} d\epsilon' a^\dagger_{\epsilon o\mu} \frac{1}{2} \vec{\sigma}_{\mu\mu'} a_{\epsilon' o\mu'} \cdot (\vec{S}_1 + \vec{S}_2) \quad (3)$$

$$+ \rho J_m \int_{-1}^{1} d\epsilon \int_{-1}^{1} d\epsilon' \left[ a^\dagger_{\epsilon e\mu} \frac{1}{2} \vec{\sigma}_{\mu\mu'} a_{\epsilon' o\mu'} + a^\dagger_{\epsilon o\mu} \frac{1}{2} \vec{\sigma}_{\mu\mu'} a_{\epsilon' e\mu'} \right] \cdot (\vec{S}_1 - \vec{S}_2).$$

There are three interactions: even and odd, which preserve the total spin of the moments; and mixed, which intermixes singlet and triplet impurity spin states, and even and odd parity conduction electrons.

Note that there are no direct interactions between the impurities themselves in this model. There are, however, indirect interactions generated to all orders by spin-flip scattering of the conduction electrons. The leading order interaction is of the form $I(r)\vec{S}_1 \cdot \vec{S}_2$ and known as the Ruderman-Kittel-Kasuya-Yosida (RKKY) interaction. If the interactions $\rho J$ between the impurity and conduction electrons are small, the RKKY interaction $I$ can be calculated by second-order perturbation theory to give the well-known oscillatory shape. For small $r$ the interaction is ferromagnetic, and then oscillates on a scale of $k_F^{-1}$ as $r$ is increased, with an envelope for large $r$ of $1/r^3$. For the above model (2), (3) we have for small (initial) $\rho J$,

$$\frac{I_o}{D} = 2 \ln 2\rho^2(2J_m^2 - J_e^2 - J_o^2) \propto (\rho J)^2. \tag{4}$$

(Both sides of the equality are implicitly $r$-dependent.) For $\rho J$ not small, we cannot calculate the inter-impurity interaction directly. Single-impurity Kondo effects tend to renormalize $\rho J$ to larger values; the resulting breakdown of perturbation theories for both $\rho J$ and $I/D$ give rise to the complex competing energy scales that necessitate the use of Wilson's procedure for a full solution.

The final transformations of this Hamiltonian are particular to the Wilson method. Wilson's numerical renormalization group method is an exact, iterative procedure for obtaining temperature-dependent properties of a Hamiltonian. Refs. 1 and 2 describe the technique as applied to the single-impurity Kondo and Anderson models, and we will not go into details here. Briefly, energy space is logarithmically discretized, and a new set of orthonormal operators $\{f_{np\mu}\}_{p=e,o;\mu=\uparrow,\downarrow}$ are defined which are roughly local to a given region of energy space, and thus also of real space. Those with index $n = 0$ are most localized about the center of the two impurities and extend the full bandwidth in energy space. The $f_1$ are somewhat less localized, and correspond to energies closer to the Fermi level. Interactions of $f_n$ for $n$ very large involve energies very close to the Fermi energy. In terms of these operators, the Hamiltonian (2), (3) becomes

$$\frac{H}{D} = \frac{1}{2}(1 + \Lambda^{-1}) \sum_{\substack{n=0 \\ p=e,o \\ \mu=\pm 1/2}}^{\infty} \Lambda^{-n/2} \xi_n [f^\dagger_{np\mu} f_{(n+1)p\mu} + f^\dagger_{(n+1)p\mu} f_{np\mu}] + 2\frac{\tilde{H}_{int}}{D}, \tag{5}$$

$$\frac{\tilde{H}_{int}}{D} = \rho \left[ J_e f^\dagger_{0e\mu} \frac{1}{2} \vec{\sigma}_{\mu\mu'} f_{0e\mu'} + J_o f^\dagger_{0o\mu} \frac{1}{2} \vec{\sigma}_{\mu\mu'} f_{0o\mu'} \right] \cdot (\vec{S}_1 + \vec{S}_2)$$
$$+ \rho J_m \left[ f^\dagger_{0e\mu} \frac{1}{2} \vec{\sigma}_{\mu\mu'} f_{0o\mu'} + f^\dagger_{0o\mu} \frac{1}{2} \vec{\sigma}_{\mu\mu'} f_{0e\mu'} \right] \cdot (\vec{S}_1 - \vec{S}_2), \tag{6}$$

with

$$\xi_n = \frac{(1 - \Lambda^{-(n+1)})}{[(1 - \Lambda^{-(2n+1)})(1 - \Lambda^{-(2n+3)})]^{1/2}}. \tag{7}$$

$\Lambda$ is the discretization parameter, which we take to be 3.0. The point is that now the Hamiltonian can be expressed in an iterative form:

$$H/D \equiv \lim_{N\to\infty} \frac{1}{2}(1 + \Lambda^{-1})\Lambda^{-(N-1)/2} H_N \tag{8}$$

with

$$H_{N+1} = \Lambda^{1/2} H_N + \xi_N(f^\dagger_{Np\mu} f_{N+1p\mu} + f^\dagger_{N+1p\mu} f_{Np\mu})  \quad (9)$$

and

$$H_0 = \frac{4}{(1+\Lambda^{-1})} \Lambda^{-1/2} \frac{\tilde{H}_{int}}{D}.$$

(Repeated indices $p$, $\mu$ are to be summed over.) The scaling by $\Lambda^{-(N-1)/2}$ insures that energy scales will be of order one for all iterations.

The input parameters are then $\rho J_e$, $\rho J_o$, and $\rho J_m$, subject to the sum rule $J_e^2 + J_o^2 + 2J_m^2 = J^2$, and such that the initial RKKY interaction (4) corresponds to a given spacing. The ground state energies are subtracted out. $H_0$ can be diagonalized by hand, and the rest of the $H_N$ are formed and diagonalized by computer (Cray XMP). After a certain number of iterations the number of states $4^{(2N+3)}$ gets unwieldy even for a Cray, and the energy spectrum is truncated, keeping only the lowest energy several (many) hundred states. (Wilson[1] has studied this truncation as an approximation and found it to give a surprisingly accurate description of the low-energy states, provided of course that the number retained at each iteration is not too small.)

The output of such a procedure is, at each iteration, a set of energy eigenvalues and eigenvectors, with their corresponding symmetry labels. The Hamiltonian (2), (3) preserves three quantities:

a) total (impurities + electron) spin
b) total (impurities + electron) parity
c) a three-vector $\vec{j}$ which we call axial charge, a property of the electrons only.

The components of axial charge are[6]

$$j^+ = \sum_{\substack{n=0 \\ p=e,o}}^{\infty} (-1)^n f^\dagger_{np\uparrow} f^\dagger_{np\downarrow}$$

$$j^- = \sum_{\substack{n=0 \\ p=e,o}}^{\infty} (-1)^n f_{np\downarrow} f_{np\uparrow} \quad (10)$$

$$j^z = \frac{1}{2} \sum_{\substack{n=0 \\ p=e,o}}^{\infty} (f^\dagger_{np\uparrow} f_{np\uparrow} + f^\dagger_{np\downarrow} f_{np\downarrow} - 1).$$

Component $j^z$ is one half the charge $q$. The operators $j^+$, $j^-$, and $j^z$ have the commutation relations of an angular momentum and all three components commute with the Hamiltonian. Hence $\vec{j}^2$ and $j^z$ are good quantum numbers and energy eigenvalues are independent of $j^z$ value. Note that particle-hole symmetry alone would require that states of $\pm q$ ($\pm j^z$) be degenerate, but conservation of axial charge further requires that all values of $q$: $-2j$, $-2j+2$, ..., $2j-2$, $2j$ be degenerate.

We will label states by three quantum numbers in the form $(2j, 2S, p)$. Here $2j$ and $2S$ are twice the total axial charge and total spin, respectively, and $p$ is parity: $0$ = even, $1$ = odd. All three indices are hence integers, and each such state has degeneracy $d = (2j+1)(2S+1)$.

## III. LOW-TEMPERATURE FIXED POINTS

As the number of iterations increases, eventually the energy levels flow to one or the other of two stable fixed points. (For $\Lambda = 3$, typically 40 iterations were sufficient.) Since large iteration numbers correspond to low temperatures, these fixed points represent the two possible (stable) ground states. The relevant parameter separating the two stable regimes is the ratio of two initial energy scales, the RKKY interaction $I_o$ (Eq. (4)) and the single-impurity Kondo temperature $T_K \approx D\sqrt{\rho J}e^{-1/\rho J}$. A summary of results[6] is as follows.

The ground state is a singlet for all nonzero impurity separations.

i) $I_o/T_K \lesssim 2.2$ For all ferromagnetic interactions ($I_o < 0$) and moderately small antiferromagnetic ones, the ground state is that of impurity spins completely quenched by the Kondo effect, with nonzero inter-impurity spin correlations.

ii) $I_o/T_K \gtrsim 2.2$ For large antiferromagnetic initial interactions, no Kondo effect occurs. The ground state is an uncompensated singlet with strong spin correlations.

iii) $I_o/T_K \approx 2.2$ For moderate antiferromagnetic initial coupling, we flow near an unstable fixed point of complex nature. This fixed point has quite unusual properties which will be detailed below.

Even though the nature of the ground state (Kondo or not) shows an abrupt transition at finite value of the coupling, the impurity spin-spin correlation $\langle \vec{S}_1 \cdot \vec{S}_2 \rangle$ is a *continuous* and monotonically decreasing function of $I_o/T_K$, even through the unstable point. In particular, it is zero only for the case of initially independent impurities, and reaches its limiting antiferromagnetic value of $-3/4$ in the ground state only in the limit of infinite $I_o$. At the transition through the unstable fixed point the value of $\langle \vec{S}_1 \cdot \vec{S}_2 \rangle$ is near $-1/4$, and there is at most a divergence in the derivative. Both the ratio $I_o/T_K$ at which the transition occurs, $2.2 \pm .2$, and the value of the impurity spin-spin correlation function at that point, $\sim -1/4$, are remarkably constant through variations of $T_K$ of over two orders of magnitude.

In the following we will briefly review the analysis[1,2] of a set of fixed point energy levels, a process which will highlight the unusual properties of the unstable state. Typically, the determination of an effective Hamiltonian for a fixed point

## Antiferromagnetic Phase Instability

is a matter of educated guesswork. The point is to try to find a Hamiltonian which preserves all the symmetries of the original Hamiltonian, and which is exactly diagonalizable to give the same energies and eigenstates as are asymptotically approached in the numerical renormalizations. A good starting point is to assign some limiting values such as 0 or infinity to the parameters of the original Hamiltonian.

In this context, we examine the free-electron Hamiltonian, obtained by setting $J = 0$ (and hence $\tilde{H}_{int} = 0$) in the Hamiltonian (9). With $\tilde{H}_{int} = 0$ the Hamiltonian simply separates into independent even and odd channels, and we consider one such channel $p$:

$$H_{N,p}^{free} \equiv \sum_{n=0}^{N-1} \Lambda^{(N-1-n)/2} \xi_n (f_{np\mu}^\dagger f_{n+1p\mu} + f_{n+1p\mu}^\dagger f_{np\mu}), \qquad p = e \text{ or } o. \qquad (11)$$

Coefficient $\xi_n$ is defined in Eq. (7) and is of order unity for large $n$. The $\Lambda^{-n/2}\xi_n$ factor is necessary in order for $H_{N,p}^{free}$ to be a hopping (kinetic energy) in logarithmically discretized space. Hamiltonian (11) can be diagonalized exactly into $N + 1$ single particle levels (per spin) by noting that $H_{N,p}^{free} = f^+ \mathcal{H}_N^o f$, where

$$(\mathcal{H}_N^o)_{lm} = (\mathcal{H}_N^o)_{ml} = \Lambda^{(N-1-m)/2} \xi_m \delta_{l,m+1}, \qquad l, m = 0, 1, 2, \ldots N \qquad (12)$$

and "$f$" $= (f_0, f_1, \ldots f_N)$. The eigenvalues $\{\eta_i\}$ of $\mathcal{H}_N^o$ are symmetric in energy space about $\eta = 0$, due to particle-hole symmetry. For $N$ large, a limiting set of values is reached, which depend only on whether $N$ is even or odd. For $N$ even there is a zero eigenvalue $\eta_i = 0$ at $i = N/2$, while for $N$ odd there is not.

The eigenstates of the free-electron Hamiltonian $H_{N,p}^{free}$ are then obtained by variably occupying the single-particle levels. The ground state of $H_{N,p}^{free}$ consists of fully occupying (two electrons each) all negative energy single-particle levels. Hence for $N$ odd the ground state is nondegenerate, while for $N$ even there is a four-fold ground state degeneracy corresponding to the variable occupancy of the $\eta = 0$ level. To subtract out the ground state energy one simply introduces hole operators conjugate to the negative-energy electron operators and normal orders in the usual way. Particles and holes then have the same set of positive (or zero) single-particle energies, and we may write the Hamiltonian as

$$H_{N,p}^{free} = \sum_{j=1}^{N/2} \hat{\eta}_j(N)(g_{jp\mu}^\dagger g_{jp\mu} + h_{jp\mu}^\dagger h_{jp\mu}) \qquad N \text{ even}$$

$$\hat{\eta}_1^* \simeq 1.70; \; \hat{\eta}_2^* \simeq 5.20; \; \ldots$$

(Complete set of operators includes $g_{0p\mu}$ of zero energy) (13)

**TABLE 1** Listing of the first few energies and eigenstates of the free-electron Hamiltonian (11) in the limit of large particle number $N$ and $\Lambda = 3.0$. The states below are those of $H_{N,e}^{free}$. The bars over the parity labels indicate those states that have odd parity for the case of $H_{N,o}^{free}$. (Recall the labeling: (2 × total axial charge, 2 × total spin, parity with even 0 and odd 1).) The degeneracy $d$ of each level is also marked.

$H_{N,e}^{free}$

$N \to \infty$, even

$E = 0$:  $(100), (01\bar{0})$   $d = 4$

$E \simeq 1.70$:  $(100), (120), (21\bar{0}), (01\bar{0})$   $d = 16$

$N \to \infty$, odd

$E = 0$:  $(000)$   $d = 1$

$E \simeq .80$:  $(11\bar{0})$   $d = 4$

$E \simeq 1.6$:  $(200), (020)$   $d = 6$

$$\sum_{j=1}^{(N+1)/2} \eta_j(N)(g_{jp\mu}^\dagger g_{jp\mu} + h_{jp\mu}^\dagger h_{jp\mu}) \qquad N \text{ odd}$$

$$\eta_1^* \simeq 0.80;\ \eta_2^* \simeq 3.00;\ \eta_3^* \simeq 9.00;\ \ldots$$

Here $\eta_i^* = \lim_{N \to \infty} \eta_i(N)$. (Likewise for $\hat{\eta}$.) The numerical values given above for the $\eta$ are for $\Lambda = 3.0$. Operator $g^\dagger$ creates a particle, and $h_{jp\mu}^\dagger$ creates a hole of parity $p$ and spin $-\mu$. The first few eigenlevels of the large-iteration-$N$ free-electron Hamiltonian for a single parity channel are accordingly as in Table 1.

We now return to the full interacting Hamiltonian with $J \neq 0$. Like the above eigenstates of a "chain" of free electrons, which depend on whether there is an even or an odd number of sites, the output of the computer diagonalization of the Hamiltonian (9) also alternates according to even or odd iteration number. In Figure 1 we show the energy flows as a function of odd iteration number for a typical value of strong antiferromagnetic RKKY coupling, $I_o/T_K \simeq 4.0$. Only the first few levels are shown. In the limit of large iterations, the energies reach plateaus, indicative of a stable fixed point. Comparison of the energy levels and eigenstates with those of the $N$ = odd free-electron Hamiltonians above reveals that all states are identical (including higher energy states not shown in the figure) to those of the sum $H_{N,e}^{free} + H_{N,o}^{free}$ ($N$ odd). (To verify, recall that both axial charge

## Antiferromagnetic Phase Instability

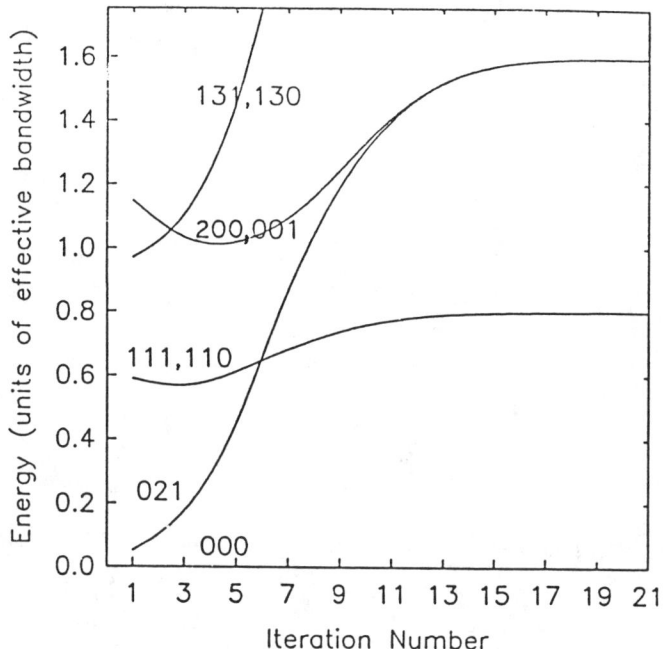

**FIGURE 1** Energies (first few levels) vs. odd iteration number for strong antiferromagnetic coupling $I_o/T_K \simeq 4.0$. [At energy $\approx 1.6$ there are a number of levels that are not illustrated for clarity. Besides (021), (200) and (001), there are also (201), (221), (020)(twice), and a second (200).] The stable fixed point, reached at around $N \approx 15$, is that of free electrons, no impurities. For this figure and also Figures 2 and 3, $\rho J = 0.25$, $\Lambda = 3.0$.

and spin must be added as vectors; e.g. (110) + (111) = (001), (021), (201), (221).) Likewise even iteration eigenvalues, not shown, are identical to those of $H_{N,e}^{free} + H_{N,o}^{free}$ with $N \to \infty$, even.

We conclude that the fixed point for $I_o/T_K \simeq 4.0$ is that of a set of free even and free odd electrons, with no remaining impurity degrees of freedom. That is, as far as the low-temperature excitations are concerned, the renormalized impurity is in a singlet ground state. One may further define asymptotic scattering states from the one-particle levels. Since the number of levels needed to describe the interacting (low-temperature) result is the same for both even and odd channels as the noninteracting case, the phase shifts for both even and odd channels are zero from the Friedel sum rule.

The flows for a typical Kondo-type fixed point are shown in Figure 2, for $I_o/T_K \simeq 0.31$, odd iterations. As in Fig. 1, only the first few levels are shown. However, one can as above get an exact match to energies and eigenstates, if one uses those of $H_{N,e}^{free} + H_{N,o}^{free}$ for *even* $N$. The conclusion is that the fixed point Hamiltonian is again that of a set of free even- and odd-parity electrons, but with the important difference that the number of one-electron levels needed to describe

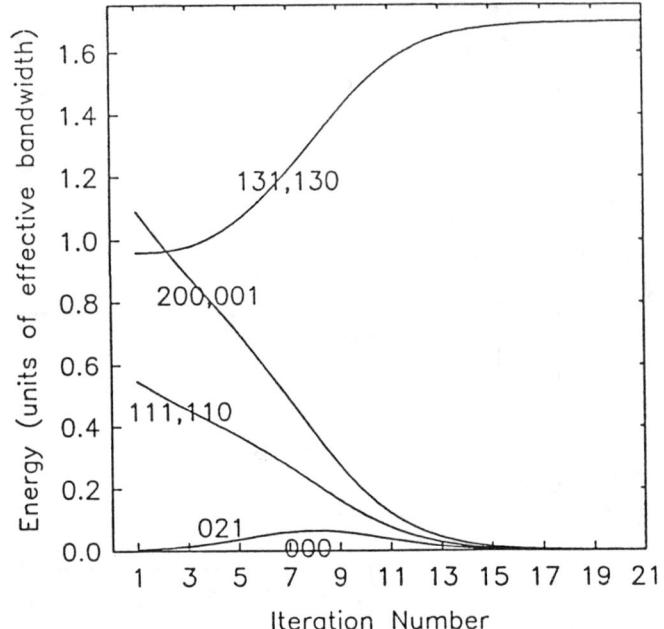

FIGURE 2 Energies (first few levels) vs. odd iteration number for weak antiferromagnetic RKKY $I_o/T_K \simeq 0.31$. The degeneracy at energy $\approx 1.7$ is 128 and only a representative pair of levels is shown. The stable fixed point is reached at iteration $N \approx 15$, and is characterized by a shift of the levels by one iteration from that of free electrons, Fig. 1. (That is, if we had plotted even iterations rather than odd in Fig. 1, the asymptotic levels would look exactly like those of Fig. 2.) This "capture" of one electron each in the even and in the odd parity channels to quench the impurity moments indicates a Kondo effect.

the interacting results are one less for both the even and the odd channels than would describe the noninteracting state. By Friedel's sum rule we obtain a phase shift of $\pi/2$ for both the even and the odd channels. Note that the results of Fig. 2 could not be described by a phase shift of less than $\pi/2$ in one channel and more than $\pi/2$ in the other, since this would produce energies which differed from the free-electron values of Table 1, and pairwise sums would not be able to replicate the unshifted totals of Fig. 2. Again the impurity degrees of freedom have disappeared from the problem, but at the expense of a phase shift at the Fermi level. This is just a rephrasing of the Kondo effect, and hence our classification of all such flows as Kondo-type fixed points.

All energy levels eventually flow to one or the other of the two fixed points that we have illustrated above. The choice of final fixed point and also the rate of approach to that fixed point are determined by $I_o/T_K$. For $I_o/T_K$ arbitrarily close to a critical value ($\approx 2.2 \pm .2$), however, the energies can be made to flow arbitrarily near an unstable fixed point which we illustrate in Figure 3. In Fig. 3

## Antiferromagnetic Phase Instability

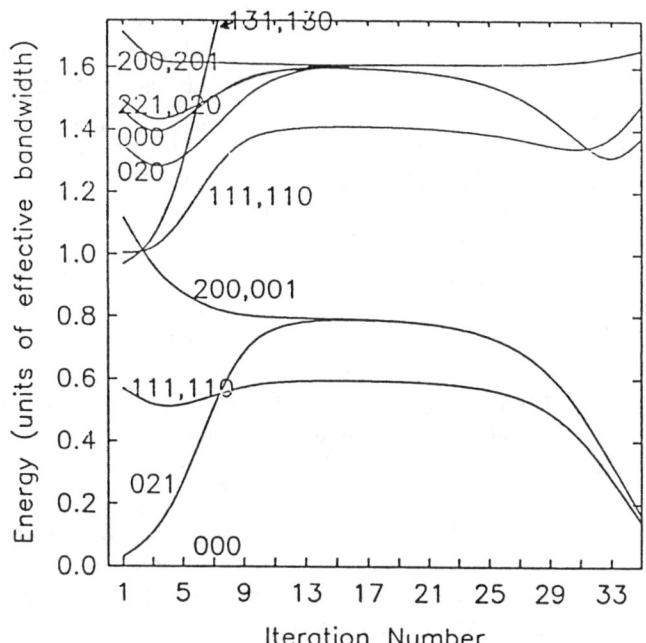

FIGURE 3  Iterations ~9 to ~29 illustrate the unstable fixed point, energy flows vs. iteration number. All levels at $E = 0$, ~.8, and ~1.6 are shown. (The states (131) and (130) are flowing to higher energy and do not belong in this grouping.) $I_o/T_K \simeq 2.4$. Odd iterations only are plotted, but during iterations 9–29, even iterations are virtually identical. The levels cannot be fit by any single-particle free-electron Hamiltonian as can those of Figs. 1 and 2.

a value of $I_o/T_K \simeq 2.4$ gives virtually constant energy levels from iterations ~9 to ~29. Around iteration 29 the flows in Fig. 3 happen to tend toward the Kondo fixed point: since the unstable fixed point occurs for a finite (probably irrational) value of the coupling parameters, it is in practice only possible to choose initial parameters such that the flows are asymptotically close to the unstable fixed point. We show only odd iterations, but during iterations 9 to 29 the energy levels and states of even iterations are virtually identical. All levels at energies 0, $\simeq .8$, and $\simeq 1.6$ are shown.

Note that the pattern of degeneracies cannot be explained by a set of free electron single particle levels, even with perhaps one or more associated zero-energy impurity degrees of freedom. The ground state is nondegenerate, and so any impurity degrees of freedom must be of finite energy. The next energy level up, at $E \approx .8$, is 15-fold degenerate, and the next, at $E \approx 1.6$, has degeneracy 30. (The deviations of (111), (110) from the other levels may be due to the presence of a marginal operator and may not be significant. If one considers the levels to have degeneracies 7, 8, and 8, 22, the following argument still holds.) These degeneracies can not be represented by single occupancy of a set of levels

$\{g_1^\dagger; h_1^\dagger; g_2^\dagger; h_2^\dagger; \ldots\}$, say, because of particle-hole and spin symmetry; for every particle operator there must be a hole operator, and for every up there must be a down. Hence even for a single parity channel the number of states at the single-particle level must be a multiple of 4. It must also be true that the fixed point is not a simple sum of even parity plus odd parity fixed points, as all the other fixed points are, since the one-channel Kondo Hamiltonian has no fixed points which do not show alternation between even and odd iterations. In short, the fixed point is quite complex in structure.

Because we cannot map the interacting case of the unstable state onto a set of free particles, there is no way to define phase shifts for this state, nor can we easily describe the renormalized state of the impurities. What is desired in an effective Hamiltonian a) which is expressed in terms of the original set of degrees of freedom, i.e., the "chain" operators $f_0, f_1, \ldots$, b) which preserves all the symmetries of the original Hamiltonian; c) which reproduces all the energy levels and quantum labels; and d) shows repetition at every iteration, not every other. Scaling the parameters of the original Hamiltonian to special values to produce free-electron operators obviously does not work, and the question remains what, if any, effective Hamiltonian to use.

It is of interest to compare the above behavior with that of another Kondo system[7,8] already known to show a fixed point at finite value of coupling constants. The system is that of an $n$-channel single-impurity spin $S$ Kondo problem with $n > 2S$. At its simplest level one considers the Hamiltonian

$$H' = H_{\text{K.E.}} + J_1 \vec{s}_{c,1} \cdot \vec{S} + J_2 \vec{s}_{c,2} \cdot \vec{S} \qquad (14)$$

with $\vec{s}_{c,1}$, $\vec{s}_{c,2}$ and $\vec{S}$ all of spin $\frac{1}{2}$. Nozières and Blandin[7] and Cragg, Lloyd, and Nozières[8] have studied this problem both analytically and via Wilson numerical renormalization group techniques, with the following results. For $J_1 = J_2 = J$ there is an attractive fixed point at finite $J \approx 2$. That is, both $J = 0$ and $J = \infty$ are unstable, and flows are toward the fixed point that is attained most rapidly for initial $J \approx 2$. For $J_1 < J_2$ the fixed point becomes unstable, and behavior is dominated by $J_2$. The salient feature of these results is that the fixed point at $J \approx 2$ also does not show any even-odd alteration, and hence does not correspond to the filling of any asymptotic one-particle levels of free electrons. No effective (single-particle) Hamiltonian can be used to fit all of the levels, and correspondingly no phase shifts can be assigned to the channels, either.

However, the structure of the dressed impurity at the finite $J$ fixed point can be determined by analysis[7,8] of the instabilities at $J = 0$ and $J = \infty$. At each stage of the iterations, the renormalized impurity is made up of an inner spin-$\frac{1}{2}$ object, bound antiferromagnetically to one electron from each channel, forming a new spin-$\frac{1}{2}$ center. This structure, duplicated at every iteration, extends outwards to the size of the system. Because of the logarithmic discretization of the Wilson procedure, the distance of each layer of electron pairs from the original spin center increases geometrically, with binding energy inversely proportional to the radius. The infinite size of the dressed impurity necessarily implies both the impossibility of defining asymptotic scattering states with respect to it and the lack of an odd-even iterations effect. [As a final note, the $N$-channel single-impurity Kondo problem was solved exactly via Bethe Ansatz by Tsvelick and Wiegmann[9] and Andrei and Destri,[10] who determined the scaling behavior at low temperatures and calculated critical exponents. These calculations did not modify the picture of the stable-$J$ fixed point described above.]

Unfortunately for analysis of the anomalous fixed point of the two-impurity Kondo problem, it is not clear that there are many useful parallels between the two Hamiltonians (3) and (14). In particular, the Hamiltonian (3) does not display the imbalance between number of scattering channels and size of impurity spin characteristic of the Cragg, Lloyd, and Nozières case above. This imbalance made both $J = 0$ and $J = \infty$ fixed points unstable and implied a stable fixed point between them. In our case, we have $n = 2S = 2$, and there are two strong coupling fixed points which are both stable. Unlike the case above, in which deviations from the well-characterized unstable points could be used to describe the structure of the stable point to which they flowed, we have the reverse case: two known stable points, with an unknown unstable point "between" them. One parallel may remain, however, and that is the lack of an odd-even effect, implying interactions which involve all the degrees of freedom of the system, and not just those closest to the impurity center. This may indicate interesting long-range (antiferromagnetic) correlations, an implication further supported by the facts of the next section.

## IV. QUANTIZING FINITE-ITERATION DEVIATIONS FROM THE STABLE FIXED POINTS

Once one has an effective Hamiltonian describing a ($N = \infty$, $T = 0$) fixed point, one can calculate deviations about the fixed point to leading order in temperature. The result is a type of Fermi liquid which enables us to calculate thermodynamic properties of the system at low temperatures. We will not go into the details of the derivation here. It turns out that there are six operators to leading order for the Kondo-type and antiferromagnetic fixed points which preserve the symmetries (parity, spin, and axial charge) of the original Hamiltonian. (If just particle-hole symmetry is considered, there are seven.) The full expansion is presented in Refs. 5 and 6. For the simplified case of $J_e = J_o$ (in Hamiltonian (3)), the unstable fixed point is still displayed, and the effective Hamiltonian reduces to just three parameters:

$$H^{eff}_{(\mathrm{AFM})} = H^{free}_e + H^{free}_o + \Delta H^{eff} \Lambda^{(N-1)/2} + O(\Lambda^{-(N-1)}) \quad (15)$$

$$\Delta H^{eff} = -t \sum_{i=1,2} (f^\dagger_{0i\mu} f_{1i\mu} + f^\dagger_{1i\mu} f_{0i\mu}) + U \sum_{i=1,2} (n_{0i} - 1)^2 + 4J_{12} \vec{s}_1 \cdot \vec{s}_2. \quad (16)$$

Here $n_{0i} \equiv f^\dagger_{0i\uparrow} f_{0i\uparrow} + f^\dagger_{0i\downarrow} f_{0i\downarrow}$ and $\vec{s}_i = f^\dagger_{0i\mu} \frac{1}{2} \vec{\sigma}_{\mu\mu'} f_{0i\mu'}$. For clarity we have reexpressed the operators $f$ in terms of impurity sites 1 and 2, rather than even and odd: $f_{i=1} \equiv \frac{1}{\sqrt{2}} (f_e + f_o); f_{i=2} \equiv \frac{1}{\sqrt{2}} (f_e - f_o)$. When the operators $f_o, f_1$ are expanded in terms of the free-electron operators $g$ and $h$ which diagonalize $H^{free}$ (cf. Eq. (13) and also Ref. 2), the terms in $\Delta H^{eff}$ are all of order $\Lambda^{-(N-1)}$, revealing the leading correction to $H^{eff}_{(AFM)}$ to be of order $\Lambda^{-(N-1)/2}$, an irrelevant operator. The fixed point is thus stable.

The $t$ term represents a hopping and can be considered a kinetic energy. One-fourth of $(n_{0i} - 1)^2$ is the square of the charge, making the $U$ term a Coulomb repulsion. Both the $U$ and $t$ terms are of the type used to describe the single-impurity Kondo problem. The interaction $J_{12}$ between the spins of the quasiparticles at sites 1 and 2 is specific to the two-impurity Hamiltonian. For the case of antiferromagnetic RKKY, $t$, $U$, and $J_{12}$ are all positive.

The coefficient $J_{12}$ is zero for independent impurities, $I_o = 0$. As the ratio $I_o/T_K$ is increased, $t$, $U$, and $J_{12}$ all increase rapidly, and ultimately diverge at the unstable fixed point. For larger $I_o/T_K$, the antiferromagnet regime, all three parameters of the effective Hamiltonian decrease, approaching zero in the limit of $I_o/T_K = \infty$. At this limiting point ($I_o = \infty$) the impurities completely disappear from the problem even at finite iterations, and the effective Hamiltonian is solely that of a set of free electrons with no residual interactions, consistent with the limiting value, $-3/4$, of the impurity spin-spin correlation function.

The single-impurity Kondo problem exhibits universal behavior with a single temperature scale $T_K$.[1] In the effective Hamiltonian, universality is displayed by a fixed ratio for $U/t$, independent of $\rho J$. For the two-impurity problem, $U/t$ also has the same fixed value, a fact which may be derived using Nozières' "weak universality" argument.[5] However, the ratio $J_{12}/t$ varies from zero at $I_o = 0$, to a maximum of $2U/t$ at the unstable fixed point, to $U/t$ in the limit of large $I_o/T_K$. The two-impurity Kondo problem thus does not display universal behavior, and the unstable fixed point can be seen as a maximum in antiferromagnetic quasiparticle spin interactions. Returning to even-odd space, we note that a value of $2U$ for $J_{12}$ at the unstable fixed point means that the coefficient $J_{eo}$ of the operator $\vec{s}_e \cdot \vec{s}_o$ becomes identically zero at the critical point. Therefore not only is the total spin $\vec{s}_e + \vec{s}_o$ conserved, but we also have an additional symmetry induced at the unstable point: the spins of even-parity and odd-parity quasiparticles $\vec{s}_e$ and $\vec{s}_o$ are conserved separately.

Using the $\Delta H^{eff}$ term in (15) as a perturbation to the free-electron effective Hamiltonian $H^{free}_e + H^{free}_o$, one can calculate thermodynamics: uniform and staggered susceptibility, and specific heat. In particular, we calculate the difference of these thermodynamic quantities from their free-electron values to get the contribution of the impurities. To leading order at low temperatures, the susceptibilities $\chi$ and $\chi_s$ are independent of temperature, and the specific heat is linear $C = \gamma T$. A full presentation of these quantities as a function of $I_o/T_K$ is given in Ref. 6. Specifically for the case of antiferromagnetic $I_o$, we find that the uniform susceptibility decreases apparently monotonically as $I_o/T_K$ is increased. There may be a discontinuity at the unstable fixed point (the error bars on $\chi$ grow too large to tell), but no divergence. In contrast, the coefficient of specific heat $\gamma$ and the staggered susceptibility $\chi_s$ increase as $I_o/T_K$ increases, and both diverge sharply at the unstable fixed point. The divergence as a function of $I_o/T_K$ follows a power law, $\chi_s, \gamma \sim [(I_o/T_K)_c - (I_o/T_K)]^{-2}$. The error bars on the exponent are around 10%, and $(I_o/T_K)_c$ is the previously quoted value $2.2 \pm .2$. Unfortunately, accuracy considerations currently prohibit the calculation of exponents for the divergence as a function of temperature for fixed $I_o/T_K$. Past the unstable point, for large $I_o/T_K$ the total thermodynamic quantities approach those of a set of noninteracting free electrons, and hence the differences $\chi$, $\chi_s$, and $\gamma$ asymptotically approach zero.

## V. CONCLUSIONS

We have summarized those properties of the two-impurity Kondo Hamiltonian that have been obtained using techniques of the numerical renormalization group, for the case of antiferromagnetic RKKY coupling between the moments. The salient feature of these results is an unstable fixed point at finite value of the ratio $I_o/T_K$, where $I_o$ is the initial (high temperature) value of the RKKY coupling (Eq. (4)), and $T_K$ is the single-impurity Kondo temperature. Of note is the fact that neither $I_o$ nor $T_K$ are parameters of the original Hamiltonian (2), (3), and moreover that $T_K$ is a many-body interaction energy describing low-temperature properties of the independent-impurity system.

By examining the energy levels of the unstable fixed point, one determines that they cannot be fit by a model of filled single-particle levels. The lack of an odd-even iterations alteration of the levels means that one cannot define asymptotic scattering states and thus phase shifts for the unstable state, which in turn implies long-range correlations about the impurity center. This concept of many degrees of freedom being involved in the unstable state is further supported by the calculation of a diverging specific heat. That the correlations are strongly antiferromagnetic is indicated by a diverging staggered susceptibility and by an effective Hamiltonian whose coefficient of inter-impurity quasiparticle spin interactions has a relative maximum. The value of the maximum means that an additional symmetry is induced at the unstable point: that of separate conservation of the spins of even and of odd parity quasiparticles.

Added to these facts is the impurity spin-spin correlation function, which is approximately $-1/4$ at the unstable point, corresponding to equal weighting of singlet and triplet of the impurity. It is as if the unstable state is a linear combination of two singlets: one a Kondo-compensated impurity triplet and the other an impurity singlet. Unfortunately this picture violates charge conservation. The alternate picture, that of a simple linear combination of impurity spin triplet and spin singlet states to form a sort of Nèel state, violates conservation of spin.

It seems clear that the structure of the unstable state is quite complex and that a full explanation will require consideration of the many-body manifold of all the conduction electron states. What is not completely clear at this point is the role of various symmetries in the formation of the instability. A preliminary study[11] of a related model using a different technique revealed a phase transition but no divergences. This model in particular did not have particle-hole symmetry. Preventing a strict comparison, however, are other differences. The calculation, an auxiliary boson, large-$N$ expansion, was done in the limit of infinite conduction electron and impurity spin, rather than spin-$\frac{1}{2}$. Fluctuations about the mean-field solution could give a variation in the critical behavior, as could the existence of a critical spin size greater than two.

Further characterization of the unstable fixed point would be provided by knowing the temperature-related critical exponents, for example, as well as by calculating various impurity-conduction electron correlation functions, and we are currently working on these projects. Several heavy fermion materials appear to be near an antiferromagnetic instability,[12] and it could be interesting to compare properties, such as neutron-scattering derived correlation functions, to those obtained from an analysis of a pair of impurities.

## ACKNOWLEDGEMENTS

The author wishes to thank A. J. Millis, G. Kotliar, A. Auerbach, and particularly C. M. Varma and J. W. Wilkins, for helpful discussions. Support was provided by an IBM postdoctoral fellowship, and by the National Science Foundation through Grant No. DMR 85-14638 and the Harvard Materials Research Laboratory.

## REFERENCES

1. K. G. Wilson, *Rev. Mod. Phys.* **47**, 773 (1975).
2. H. R. Krishna-murthy, J. W. Wilkins, and K. G. Wilson, *Phys. Rev. B* **21**, 1003 (1980).
3. Of note are the early thermodynamic scaling results of C. Jayaprakash, H. R. Krishna-murthy, and J. W. Wilkins, *Phys. Rev. Lett.* **47**, 737 (1981), who studied the cases $|I_o| \gg T_K$. For Monte Carlo results for two Anderson impurities see J. E. Hirsch and R. M. Fye, *Phys. Rev. Lett.* **56**, 2521 (1986); R. M. Fye, J. E. Hirsch, and D. J. Scalapino, *Phys. Rev. B* **35**, 4901 (1987).

4. B. A Jones, Ph.D. Thesis, Cornell University, 1987.
5. B. A. Jones and C. M. Varma, *Phys. Rev. Lett.* **58**, 843 (1987).
6. B. A. Jones, C. M. Varma, and J. W. Wilkins, *Phys. Rev. Lett.* **61**, 125 (1988).
7. P. Nozières and A. Blandin, *J. Physique* **41**, 193 (1980).
8. D. M. Cragg, P. Lloyd, and P. Nozières, *J. Phys. C* **13**, 803 (1980).
9. A. M. Tsvelick and P. B. Wiegmann, *Z. Physik B* **54**, 201 (1984).
10. N. Andrei and C. Destri, *Phys. Rev. Lett.* **52**, 364 (1984).
11. B. A. Jones, G. Kotliar, and A. J. Millis, *Phys. Rev. B* **39**, (1989) (Feb. 15 issue); See also A. J. Millis article in this workshop proceedings.
12. A recent neutron scattering study is, e.g., G. Aeppli, E. Bucher, C. Broholm, J. K. Kjems, J. Baumann, and J. Hufnagl, *Phys. Rev. Lett.* **60**, 615 (1988).

**Mehran Kardar**
Department of Physics
MIT
Cambridge, MA 02139

# Field Theories for Elasticity of Tethered Networks

Tethered networks are large macromolecules with regular ($D$-dimensional) connectivity, embedded in a $d$-dimensional space. A "Landau-Ginzburg" field theory is proposed to describe statistical mechanics of such structures. Effective elasticity and rigidity terms emerge naturally as a consequence of connectivity, while longer-range interactions are also included. A combined Mean-Field/Flory treatment reveals crumpled, rigid, and compact phases. Fluctuations (as explored by various renormalization-group schemes) play a crucial role in each of these phases, and at phase transitions.

## I. MODEL

### 1.1 Motivation

Recently much attention has been devoted to understanding the statistical mechanics of surfaces.[1] The interest spans particle physics (string and gauge theories), condensed matter physics (microemulsions, lamellar phases), and biology (membranes). However, the statistical mechanics of two-dimensional manifolds has turned out to be considerably more complex than that of one-dimensional objects[2] (polymers). Indeed a variety of different macroscopic behaviors can emerge from the choice of different starting microscopic surface models. In view of this complication, the microscopic model under consideration has to be carefully defined. Recently we introduced[3] a surface model composed of permanently connected particles. This can be regarded as the simplest generalization of polymers. Indeed, one can further generalize by considering $D$-dimensionally connected structures.[4] Such "macromolecules" will be referred to as "tethered networks".

The constraint of fixed connectivity is easily implemented by assigning a $d$-dimensional coordinate $\vec{r}_i$ to each particle, and an energy cost $V(|\vec{r}_i - \vec{r}_j|)$ which diverges for large separations. (Note that $d = 3$ for any real problem, but it is convenient to consider the general case.) Such a description clearly applies to covalently bonded atoms at low temperatures. (There can, in addition, be many body interactions.) For a statistical description of a large network the positions $\vec{r}_i$ of individual particles are no longer appropriate. Instead, we consider a coarse-grained average $\vec{r}(\mathbf{x})$; where a *continuous* coordinate $\mathbf{x}$ replaces the discrete index $i$ for the *internal* position of the particle in the network, and the external coordinate $\vec{r}$ is the average of the positions of particles in the vicinity of $\mathbf{x}$ on the network.

All information about connectivity now resides in the topology of the domain $\Omega$, which is the collection of all possible coordinates $\mathbf{x}$. Here, to proceed further, we make the assumption that $\Omega$ is a regular $D$-dimensional space of typical linear size $L$. (The volume $L^D$ is proportional to $N$, the total number of atoms in the molecule.) For example, $D = 1$ corresponds to a linear macromolecule, while $D = 2$ refers to a planar structure such as a fisherman's net. Gels can be regarded as $D = 3$ networks, while non-integer $D$ may approximate fractal structures.

The interparticle interactions can be separated into two groups:[4] (i) Corresponding to strong bonds between neighboring particles on the network (these are responsible for the primary conformations of biological molecules); and (ii) Weaker hydrogen bonds, van der Waals, and hardcore interactions between particles that are far away along the internal coordinate $\mathbf{x}$, but are actually close in real space (responsible for secondary and tertiary conformations). The role and consequences of each type in a statistical description is now described separately.

## 1.2 Internal Connectivity

First consider the strong bonds holding the network together, and ignore interactions between particles not close in internal space. In the terminology of polymers,[2] these are "phantom networks." The probability of a particular configuration $\{\vec{r}(\mathbf{x})\}$ is given by the Boltzmann weight $P(\{\vec{r}(\mathbf{x})\} \propto \exp[-\beta H\{\vec{r}(\mathbf{x})\}]$. Rather than calculating the energy function $\beta H$ from first principles, we deduce its form from symmetry considerations. This is similar in spirit to the Landau-Ginzburg approach to superconductivity, and phase transitions.[5] The basic symmetries in external space are invariance under translations and rotations. The former implies that $\beta H$ can only depend on derivatives such as $\partial_\alpha \vec{r} \equiv \partial \vec{r}(\mathbf{x})/\partial x_\alpha$ and $\partial_\alpha \partial_\beta \vec{r}$; while

# Field Theories for Elasticity of Tethered Networks

the latter allows only scalar products such as $\partial_\alpha \vec{r} \cdot \partial_\beta \vec{r}$. For a uniform *and isotropic* network we can then expand $\beta H$ in powers of $\partial_\alpha \vec{r}$

$$\beta H_1 = \int d^D x \left[ \frac{t}{2} (\partial_\alpha r_i)^2 + u(\partial_\alpha r_i \partial_\beta r_i)^2 + v(\partial_\alpha r_i \partial_\alpha r_i)^2 + \frac{\kappa}{2} (\partial_\alpha^2 r_i)^2 + \cdots \right], \quad (1)$$

where $\alpha = 1, 2, \ldots, D$; $i = 1, 2, \ldots, d$; and summation convention is used.[4]

The terms in Eq. (1) have simple physical interpretations: $t$ represents a Hookian elasticity, while $u$ and $v$ are anharmonic elastic terms; $\kappa$ corresponds to a rigidity that arises from bond-bending forces. Thus elasticity and rigidity emerge naturally as macroscopic manifestations of internal connectivity. For polymers, entropy-generated elasticity is well known and easily proved by the central limit theorem.[2] Its presence for $D = 2$ has been shown from simulations, and can be seen in all $D$ from an approximate Migdal-Kadanoff rescaling.[3] In fact Eq. (1) describes a variety of macroscopic behaviors,[4] in the same way that Landau theories of magnetism describe ordered and disordered phases.[5] A detailed description of these phases will be given after inclusion of long-range interactions. It is clear, for example, that for $t < 0$ and $(u, v) > 0$, the tangents $\partial_\alpha \vec{r}$ obtain a non-zero expectation value. This broken symmetry state is in fact a rigid phase.[6]

## 1.3 Long-Range Interactions

The entropy generated elastic energy is minimized when the network has a small size $R$ in external space.[3] Consequently particles that are far apart along the backbone of the manifold are brought to close physical proximity and their interactions can no longer be ignored. A general two body term is described by $\int d^D x d^D y U(|\vec{r}(\mathbf{x}) - \vec{r}(\mathbf{y})|)$. At temperatures higher than typical van der Waals energies, particles only feel the hardcore repulsion which can be described by $U \cong b(T) \delta^d[\vec{r}(\mathbf{x}) - \vec{r}(\mathbf{y})]$. As in polymers the "excluded volume" $b(T)$ is related to the second virial coefficient in a solution of such networks, and may change sign as temperature is reduced, or quality of the solvent is changed.[2] When this happens, higher order terms are necessary to prevent collapse to very high densities. With this in mind we approximate the effects of these remote interactions by an energy

$$\beta H_2 = b \int d^D x_1 d^D x_2 \delta^d[\vec{r}(\mathbf{x}_1) - \vec{r}(\mathbf{x}_2)]$$

$$+ c \int d^D x_1 d^D x_2 d^D x_3 \delta^d[\vec{r}(\mathbf{x}_1) - \vec{r}(\mathbf{x}_2)] \delta^d[\vec{r}(\mathbf{x}_2) - \vec{r}(\mathbf{x}_3)]. \quad (2)$$

The probability for a configuration $\{\vec{r}(\mathbf{x})\}$ is now controlled by a Hamiltonian $\beta H = \beta H_1 + \beta H_2$; including both neighboring and remote interactions. The terms in the Hamiltonian were written down on the basis of symmetry, and hence the parameters $t$, $u$, $v$, $\kappa$, $b$, and $c$ are all expected to be functions of temperature, microscopic parameters, solution pH and so on. As in usual Landau theories the signs of these effective parameters determine the macroscopic phase of the network.[5] Such analogies are formally correct only for $L, N \to \infty$; but for large $N$ similar behavior is expected up to some "finite-size rounding" effects which will also be discussed.

## II. MEAN-FIELD/FLORY TREATMENT

Consider a network of internal linear size $L$ (or mass $L^D$), embedded in $d$-dimensions. The scaling of the actual network size $R$ in external space with $L$ characterizes the macroscopic "phase" of the system. This dependence can be estimated by comparing typical values for the various terms in $\beta H_1$ and $\beta H_2$. These estimates[6] are summarized by a free energy

$$2\beta F/D \approx \kappa R^2 L^{D-4} + tR^2 L^{D-2} + wR^4 L^{D-4} + bL^{2D}/DR^d + cL^{3D}/DR^{2d}, \quad (3)$$

where the $\beta H_1$ terms are evaluated approximately by replacing $\partial_\alpha r_i$ with $R/L$, and $w = u + Dv$. The terms in $\beta H_2$ are calculated in the usual Flory approximation.[2] As $R/L$ is typically small, the expansion in Eq. (3) is justified, and only the most important terms need to be kept. For example the rigidity term is always smaller than the elastic terms and can be neglected. Since $b$ and $t$ can change sign with temperature,[2,6] the other terms with $c, w > 0$ are necessary to ensure stability.

The optimal radius is obtained by minimizing Eq. (3) with respect to $R$ for fixed $L$. At sufficiently high temperatures both $t$ and $b$ are positive and these two terms asymptotically dominate the rest. Minimization leads to the generalized Flory expression[6,7]

$$R \sim \left(\frac{b}{t}\right)^{\frac{1}{d+2}} L^{\frac{D+2}{d+2}}, \quad (4)$$

i.e. balancing the self-avoiding repulsion with the elastic attraction leads to a *crumpled* network with a non-trivial fractal dimension. The Flory exponent $\nu_F = (D + 2)/(d + 2)$, although not exact, is quite accurate for polymers ($D = 1$)[2]; and $\nu_F = 4/5$ is quite close to the result obtained numerically[3] for surfaces ($D = 2$) in $d = 3$.

Eq. (4) indicates a divergence in $R$ as $t \to 0$; and indeed for $t < 0$ the anharmonic term with $wR^4 L^{D-4}$ is needed for stability. The competition between these terms leads[6] to $R \sim L|t|^{1/2}$. This clearly describes an expanded *rigid* phase with stretched bonds between particles on the network. The average bond length plays the role of an order parameter, and vanishes with the mean-field exponent $\beta = 1/2$ close to the transition ($|t| \to 0$).

At $t = 0$, the anharmonic and self-avoiding terms compete leading to[6] $R \sim L_c^{\nu_c}$ with $\nu_c = (D + 4)/(d + 4)$. The distinct scaling forms in the vicinity of $t \sim 0$ can be combined into a single homogeneous function $R \sim L_c^{\nu_c}\Psi(tL^y)$, where $\Psi(0) =$ constant, $\Psi(x) \to |x|^{\phi_\pm}$ for $x \to \infty$, $y = 2(d - D)/(d + 4)$, $\phi_- = 1/2$, and $\phi_+ = -1/(d + 2)$. Similar homogeneous functions, with the same crossover exponent $y$ can be constructed for other variables such as the free energy and heat capacity.[6] For example, approaching the transition from the crumpled side, the specific heat per particle scales anomalously, $C \sim t^{-\alpha}L^{-k}$ with $\alpha = (d + 4)/(d + 2)$ and $k = 2(d - D)/(d + 2)$. These results only hold for $t > t_x \sim L^{-y}$; and may be experimentally observable.

Again Eq. (4) predicts $R \to 0$ as $b \to 0$; and for $b < 0$ the three body interaction with $c > 0$ is necessary for stability. Consequently $R \sim \left(\dfrac{c}{b}\right)^{1/d} L^{D/d}$ which describes a *compact* structure with a density that vanishes linearly with $b$. The point $b = 0$ is referred to as a θ-point in polymer terminology,[2] and is similar to a tricritical point.[5] At this point balancing the three-body and elastic energies leads to $R \sim L^{\nu_\theta}$ with $\nu_\theta = (D + 1)/(d + 1)$. In the vicinity of $b \sim 0$ there is again a scaling form $R \sim L^{\nu_\theta}\Phi(bL^{y_\theta})$ that can be matched to the forms given for $b > 0$ and $b < 0$. There is again a specific heat anomaly approaching the θ point from the crumpled side.[2] One should note however that for $b < 0$, a solution of such networks phase separates,[3] and the results from looking at a single network are no longer meaningful.

For $b, t < 0$ there is a first order transition between the rigid and compact phases for $b^2/c \sim t^2/w$. The resulting phase diagram is indicated if Fig. (1). The three phase boundaries come together in a special multicritical point at $b = t = 0$. At this point scaling is determined by the competition of anharmonic and three body terms, leading to $R \sim L^{\nu*}$ with $\nu* = (D + 2)/(d + 2)$ which coincides with the Flory exponent $\nu_F$.

## III. FLUCTUATIONS

### 3.1 Crumpled Networks

The crumpled phase is "critical" in the sense that it has no characteristic length-scale. (Tangent-tangent correlations, for example, decay algebraically.) It can be shown that[4,8] for $d > d*(D) = 4D/(2 - D)$ self-avoidance is irrelevant and $\nu$ takes the free-field value $\nu_0 = (2 - D)/2$. For $d > d*(D)$ upon including fluctuations $R$ behaves as $R \sim b^{\psi'}t^{-\psi}L^\nu$. The exponent $\nu$ can be calculated systematically by an

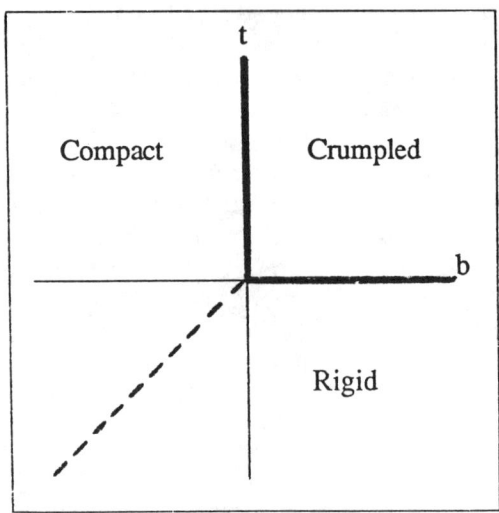

FIGURE 1 Schematic phase diagram obtained from Eq. (3).

$\tilde{\epsilon} = 4D - d(2 - D)$ expansion.[4,8] Furthermore, since there is a general scaling form

$$R^2 \sim t^{-1}L^{2-D}f(bt^{d/2}L^{\tilde{\epsilon}/2}), \tag{5}$$

with $f(x) \sim x^{2(2\nu+D-2)/\tilde{\epsilon}}$ for large $x$; the exponents $\omega$ and $\omega'$ determining critical behavior for $b, t \to 0$ can be expressed entirely in terms of $\nu$ (e.g.[6] $\omega = (2D - d\nu)/\tilde{\epsilon}$). Relaxation of these surfaces is by diffusive modes, and their critical dynamics have been described in detail elsewhere.[3]

## 3.2 The Rigid Phase

In the rigid phase there is a spontaneously broken symmetry, as the network orients itself along a particular direction in space. We should therefore consider fluctuations in the form of Goldstone modes that tend to restore the symmetry. The lowest energy modes are undulations $h_\beta$ ($\beta = D + 1, \ldots, d$) in directions perpendicular to the chosen orientation of the network. These modes are, however, coupled to phonons $u_\alpha$ ($\alpha = 1, \ldots, D$); along the stretched network backbone. Substituting $\vec{r}(\mathbf{x}) = \zeta[(x_\alpha + u_\alpha)\hat{e}_\alpha + h_\beta \hat{e}_\beta]$ in $\beta H_1$ and using $\zeta \sim R/L \sim |t|^{1/2}/w$ leads to[6]

$$\beta H_f = \int d^D x \left[ \frac{\kappa}{2}(\nabla^2 h_\beta)^2 + \mu u_{ij}^2 + \frac{\lambda}{2} u_{kk}^2 \right], \tag{6}$$

where the strain tensor is $u_{ij} = \frac{1}{2}[\partial_i u_j + \partial_j u_i + \partial_i h_\beta \partial_j h_\beta]$, and the elastic constants are $\mu = 4u\zeta^4$ and $\lambda = 8v\zeta^4$. Eq. (6) generalizes the expression for the bending of plates.[9] Note that the elastic constants $\mu$ and $\lambda$ vanish at the transition.

Naively one expects that the undulation modes will restore the full symmetry for $D \leq 2$, and that the rigid phase is unstable for $D = 2$. (Indeed this mechanism is responsible for absence of rigid-rod polymers). However, the non-linear coupling of undulation and phonon modes tends to stiffen the flat surface at large distances. This is seen both in a mode-coupling approximation,[10] and in a rigorous $\epsilon = 4 - D$ expansion.[11] A flat tethered surface has also been observed upon inclusion of bending rigidity in the numerical simulations.[12]

### 3.3 The Compact Phase

Minimizing the free energy in Eq. (3) for $b < 0$ led to a compact phase. This phase can also be obtained by exerting an external pressure $p$ on the network. However, the compact equilibrium configuration may not be easily accessible dynamically, and the system may get trapped into metastable configurations. Indeed, although it is easy to compactify a string, attempts to compactify aluminum foil,[4] or to crush paper[13] lead to fractal objects with $\nu \sim 4/5$ rather than $2/3$. These results, although quite reproducible, have not been satisfactorily explained.

### 3.4 The Rigid to Crumpled Transition

The mean-field approach described in Sec. 2 ignores critical fluctuations. As the crumpling transition is approached from the flat phase, a Ginzburg criterion[6] shows that mean-field theory breaks down for $t^2(\kappa/-t)^{D/2} \lesssim 1$. If the transition remains continuous, we expect $R \sim L|t|^\beta$ with $\beta \neq 1/2$ due to fluctuations. Since self-avoidance is irrelevant in the flat phase, the exponent $\beta$ can in fact be probed systematically by $\epsilon = 4 - D$ expansions.[5] To the lowest order in $\epsilon = 4 - D$, the transition is in fact driven first order by fluctuations[6] for $d < d_c \approx 219$. It is however possible that a continuous transition is recovered for $\epsilon = 0(1)$ as suggested by Monte-Carlo simulations[12] in $D = 2$ and $d = 3$.

Another approach that includes fluctuations at the rigid to crumpled transition is a $1/d$ expansion (analogous to the $1/n$ expansion for $O(n)$ models[5]). Indeed this approach at $D = 2$ has suggested a continuous transition,[14] and its extension to general $D$ is quite natural.[15,16] The resulting field theory has the remarkable property of having lower critical dimension $D_l = 2 - 2/d + O(1/d^2)$, which is less than two. This theory can thus be used to systematically probe for the first time the approach to a nontrivial lower critical dimension.[16]

## IV. CONCLUSIONS

Guided by considerations of symmetry, we have constructed a field theory describing embeddings of a regular isotropic network in $d$-dimensional space. Elasticity and rigidity terms naturally appear as a result of internal connectivity; and interactions between particles remote on the network are also included. This theory is capable of describing various phases of the network: crumpled, rigid, compact, etc. The behavior of these phases and the transitions between them is described both in a mean-field/Flory approach, and upon inclusion of fluctuations. This model represents a unified approach to regular networks ($D = 1$ for polymers, $D = 2$ for tethered membranes, and $D = 3$ for gels). Inclusion of fluctuations has led us to consider some remarkable field theories. The crumpled phase is described by a theory of fractional manifolds, and perturbation theory requires calculation of manifold diagrams with similarities to string-theory diagrams. Consideration of rigid phase indicates unusual couplings between Goldstone modes, yielding anomalously scaling elastic constants. Consequently the Mermin-Wagner theorem no longer holds, and a spontaneous symmetry breaking is allowed in lower than two dimensions.

## ACKNOWLEDGMENTS

The work described here has resulted from collaborations with Y. Kantor, D. R. Nelson, and M. Paczuski. The research was supported by NSF through grant No. DMR86-20386.

## REFERENCES

1. *Statistical Mechanics of Membranes and Surfaces*, Proceedings of the Fifth Jerusalem Winter School, edited by D. R. Nelson, T. Piran, and S. Weinberg (World Scientific, Singapore, in press).
2. P. G. De Gennes, *Scaling Concepts in Polymer Physics* (Cornell University Press, Ithaca, NY, 1979).
3. Y. Kantor, M. Kardar, and D. R. Nelson, *Phys. Rev. A* **35**, 3056 (1987).
4. M. Kardar and D. R. Nelson, *Phys. Rev. A* **38**, 966 (1988).
5. K. G. Wilson and J. Kogut, *Phys. Repts* **12C**, 75 (1974).
6. M. Paczuski, M. Kardar, and D. R. Nelson, *Phys. Rev. Lett.* **60**, 2638 (1984).
7. M. E. Cates, *Phys. Lett.* **161B**, 363 (1985); *Phys. Rev. Lett.* **53**, 926 (1984).
8. J. A. Aronovitz and T. C. Lubensky, *Europhys. Lett.* **4**, 395 (1987).
9. L. D. Landau and E. M. Lifshitz, *Theory of Elasticity*, (Pergannon, 1986).
10. D. R. Nelson and L. Peliti, *J. Physique* **48**, 1085 (1987).
11. J. A. Aronovitz and T. C. Lubensky, *Phys. Rev. Lett.* **60**, 2634 (1988).
12. Y. Kantor and D. R. Nelson, *Phys. Rev. A* **36**, 4020 (1987).
13. M. A. F. Gomez and G. L. Vasconcelas, *Phys. Rev. Lett.* **60**, 237 (1988); Y. Kantor, M. Kardar, and D. R. Nelson, *Phys. Rev. Lett.* **60**, 238 (1988).
14. F. David and E. Guitter, *EuroPhys. Lett.* **5**, 709 (1988).
15. J. A. Aronovitz, L. Golubovic, and T. C. Lubensky, UPenn., preprint (1988).
16. M. Paczuski and M. Kardar, MIT preprint (1989).

**Steven Kivelson**
Department of Physics
State University of New York at Stony Brook
Stony Brook, NY 11794

# Statistics of Holons in the Quantum Hard-Core Dimer Gas

The nature of holons, and in particular, their statistics, are studied in the context of a two-dimensional quantum hard-core dimer gas. This model has been shown to embody the low-energy physics of the short-ranged resonating-valence-bond state. We find that, depending on detailed energetic considerations, the holons either bind a half flux quantum of "statistical flux," in which case they are fermions, or they are free, in which case they are bosons. The exchange energy is shown to favor a fermionic hole while the hole kinetic energy (which generally exceeds the exchange energy) favors bosonic holes. Finally, it is shown that even in the bosonic case, flux quantization is in units of $hc/2e$.

## I. INTRODUCTION

The notion[1-3] that quasiparticles with unusual quantum numbers (e.g., fractional charge) and unusual statistics (e.g., fractional statistics) can describe the low-energy excitations of two-dimensional quantum systems has played an important role in understanding the fractional quantum Hall effect,[2] and more recently has been invoked in the context of high-temperature superconductivity.[3-11] In particular, in Ref. 3 it was suggested that in a system with a short-ranged resonating-valence-bond (RVB) ground state, the quasiparticles created on doping are charge

$e$, spin-0 holons satisfying Bose statistics. While the reverse charge spin relation of the quasiparticle is unambiguous, and indeed is a rather general consequence of singlet pairing,[12] the statistics of the quasiparticles is a subtle issue. The subtlety arises because in two dimensions, statistics are a matter of convenience; the statistics of a particle can be changed at will by attaching a partial flux quantum of "statistical flux" to each particle.[1] At best, one particle choice of statistics is natural if it eliminates all long-range pieces of the particle kinetic energy. Partly for this reason, the assignment of statistics has proven to be highly controversial; in Refs. 4–8 the holons were identified as bosons, in Ref. 9 as half-fermions, and in Refs. 10 and 11 as fermions. The important point here is that even when there exists an unambiguous natural choice for the statistics of the low-energy quasiparticles, that choice can change depending on detailed energetic considerations which determine whether a particle will or will not bind a flux. Indeed, Read and Chakraborty[11] have pointed out that exactly this issue arises in the context of the spinons and holons in a short-ranged RVB state.

In this paper we will analyze a simple model of a short-ranged RVB superconductor, the quantum hard-core dimer gas,[5] in which the statistics of the holon can be understood completely. (By inference, the statistics of the spinon can be understood in the same way, although there are a variety of technical issues which make the case of the spinon more complicated.) We find that the two-dimensional dimer model has vortex excitations which carry one half flux quantum of statistical flux; the purely magnetic interactions in the model cause a bare holon to bind a vortex, thereby turning it into a fermion. These conclusions are in agreement with the results of Chakraborty and Read[11] based on an analysis of the nearest-neighbor RVB state with phases chosen to satisfy the Marshall[13] sign rule. However, we find that the holon kinetic energy causes the holon to unbind from the vortex leaving a bare (i.e., bosonic) holon as the low-energy quasiparticle. Moreover, we argue that in three dimensions and higher, where statistics are robust, the holon is always a boson. Finally, we show that the presence of vortex excitations leads to electromagnetic flux quantization in units of $hc/2e$, even when the ground state is a Bose condensate of charge $e$ bosons.

## II. THE MODEL

Recently, it has been shown that there is a class of models[14] whose ground state and low-lying excited states lie in the subspace spanned by the nearest-neighbor valence-bond states. These models are described by a Hamiltonian which is the sum of the so-called Klein Hamiltonian,[15] $H_K$, plus any of a broad class of perturbing Hamiltonians, $H'$. The ground-state manifold of $H_K$ is the subspace $\Omega_{\text{NNVB}}$ spanned by the nearest-neighbor valence-bond states. We treat $H'$ using degenerate perturbation which amounts to projecting $H'$ into $\Omega_{\text{NNVB}}$. In another paper[5] we do this for a representative $H'$ which is the sum of a nearest-neighbor antiferromagnetic Heisenberg interaction plus a holon kinetic energy; however, none of our main results depend sensitively on the exact nature of the perturbation. Since the states in $\Omega_{\text{NNVB}}$ are clearly in one-to-one correspondence with the states of a hard-core dimer gas, by simply orthogonalizing the valence-bond states the system can be mapped onto a hard-core quantum dimer gas on a lattice. Roughly, a dimer represents a valence bond. The nonorthogonality of the valence-bond states is represented *exactly* by the presence of longer-range interactions in the dimer Hamiltonian; the fact that the valence-bond states are, in a sense, nearly orthogonal is reflected in the fact that the dimer Hamiltonian $H_{\text{dim}}$ is short ranged (in the sense of exponentially falling).

The details of the derivation of the hard-core dimer model from the perturbed Klein model are discussed in Ref. 5; we sketch the procedure here. First, we must define a phase convention for valence-bond states. We have adopted the convention that a valence-bond between sites $i$ and $j$ is created by

$$b_{ij}^\dagger = \frac{1}{\sqrt{2}} (c_{i\uparrow}^\dagger c_{j\downarrow}^\dagger + c_{j\uparrow}^\dagger c_{i\downarrow}^\dagger), \tag{1}$$

where $c_{is}^\dagger$ creates an electron of spin $s$ on site $i$. A valence-bond state is created by the set of $b_{ij}^\dagger$ corresponding to a dimer configuration $c$. The valence-bond states can be orthogonalized by the method of Lowdin,[16] and we define a "dimer" state to be the member of the resulting orthonormal set corresponding to dimer configuration $c$. The matrix elements of the $H_{\text{dim}}$ are then the matrix elements of $H'$ between dimer states. We have shown that for short-ranged perturbations, only short-ranged terms in $H_{\text{dim}}$ are important; so for simplicity, we will consider a model consisting of only the shortest-range term of each type. We do not expect other terms in the dimer Hamiltonian to affect our conclusions. Therefore, we consider the model represented below:

$$H_{\text{dim}} = H_J + H_{J'} + H_T, \tag{2}$$

$$H_J = J \sum \left[ |\mathord{=}\rangle \langle \mathord{||}| + \text{H.c.} \right]$$

$$+ V \sum \left[ |\mathord{=}\rangle \langle \mathord{=}| + |\mathord{||}\rangle \langle \mathord{||}| \right], \tag{3}$$

$$H_{J'} = J' \sum \left[ |\mathord{\rlap{\raise1pt\hbox{\scriptsize-}}{\cdot}}\rangle \langle \mathord{\raise1pt\hbox{\scriptsize-}\llap{\cdot}}| + \text{H.c.} \right], \tag{4}$$

$$H_T = -t \sum \left[ |\mathord{\cdot\,\mathord{!}}\rangle \langle \mathord{\_\!\_}| + \text{H.c.} \right] + V_h \sum \left[ |\mathord{\cdot\,\cdot}\rangle \langle \mathord{\cdot\,\cdot}| \right], \tag{5}$$

where the sums in Eq. (3) run over all plaquettes, the sum in Eq. (4) over the neighborhood of each hole, and the first sum in Eq. (5) runs over all triplets of nearest-neighbor sites, while the second sum is over nearest-neighbor pairs of sites. The matrix elements of a bra containing a particular local dimer configuration with an arbitrary dimer state ket is zero unless the two dimer configurations match, in which case it is one. In Eqs. (3)–(5), a bar represents a dimer and a dot represents an empty site or bare holon. The first two terms, $H_J$ and $H_{J'}$, are purely magnetic in origin and so are proportional to the exchange coupling; the final term $H_T$ arises from electron hopping between sites. For a nearest-neighbor antiferromagnetic exchange interaction, we expect $J$ and $J'$ both to be positive; we shall see that this favors a ground-state superposition of dimer states with the relative phases that are expected on the basis of the Marshall rule.[11,13] For a sufficiently frustrated magnetic system it is possible for the ground state to violate the Marshall rule, i.e., it is also reasonable to consider the model with $J$ and $J'$ negative. In contrast, the pure hopping Hamiltonian always favors a $\mathbf{k} = 0$ state, so $t$ is always expected to be positive. Thus the hole kinetic energy favors a totally symmetric superposition of dimer states and for positive $J$ and $J'$, the doped system is somewhat frustrated. It is useful to include the longer-range interaction $J'$ in the Hamiltonian even though we except it to be small compared to $J$, since without it the purely magnetic part of the Hamiltonian has a conserved winding number [defined in Ref. 4(b)] about each hole and thus $H_J$ is block diagonal. The "pinwheel operator" in Eq. (4) removes this unphysical aspect of the static-hole model. The various terms in the dimer Hamiltonian can be interpreted as a pure dimer kinetic energy ($J$), a dimer potential energy ($V$), a holon kinetic energy ($t$), and a holon-holon repulsion ($V_h$).

## III. STATISTICS OF THE HOLONS IN THE 2D MODEL

An arbitrary state $|\psi\rangle$ in the dimer Hilbert space can be represented as a liner combination of dimer states $|c\rangle$ with specified amplitude $A_c$ and phase $\theta_c$,

$$|\psi\rangle = \sum_c A_c e^{i\theta_c} |c\rangle, \tag{6}$$

where $c$ specifies a dimer configuration (including, implicitly, the locations of the holes). For a finite-size system, if there is a choice of phases which makes the expectation value of all off-diagonal matrix elements of the Hamiltonian negative (i.e., if the system is unfrustrated), then the ground-state wave function has those phases. In many cases, it is convenient to define a phase field $a(l)$ which lives on the links $l$ of the lattice such that

$$\theta_c = \sum_{l \in c} a(l). \tag{7}$$

Notice that in the absence of the hole kinetic energy there is a gauge invariance, analogous to the $U(1)$ gauge symmetry discussed by Baskaran and Anderson[17] for the Heisenberg model, such that all energies are invariant under

$$a(l) \to a(l) + \chi(\mathbf{R}) + \chi(\mathbf{R}'), \tag{8}$$

where $\mathbf{R}$ and $\mathbf{R}'$ are the lattice sites on either end of link $l$, and $\chi(\mathbf{R})$ is an arbitrary function of $\mathbf{R}$. Indeed, we[18] have recently shown that the dimer model defined above is equivalent to compact lattice quantum electrodynamics (QED), and even in the presence of dynamical holes the model can be made gauge invariant by defining a new field $\phi(\mathbf{R}) = |\phi(\mathbf{R})|e^{ia(\mathbf{R})}$ associated with the holons such that

$$\theta_c = \sum_{l \in c} a(l) + \sum_{\mathbf{R} \in c} a(\mathbf{R}), \tag{9}$$

where the second sum is over all unoccupied sites (oocupied by holons); under a gauge transformation, $a$ still transforms as in Eq. (8), and

$$\phi(\mathbf{R}) \to \phi(\mathbf{R})e^{i\chi(\mathbf{R})}. \tag{10}$$

Because of the gauge invariance of the model, the relevant information about $\theta_c$ can be summarized by specifying the distribution of $a$ flux in the system; we refer to this as "statistical flux." For $J$ positive, the off-diagonal matrix elements of $H_J$ are all negative if there is a half-integer flux quantum through each plaquette. If in addition $J'$ is positive, and if we introduce an additional half flux-quantum distributed in an arbitrary fashion through the four plaquettes which surround each holon, then all the matrix elements of $H_{J'}$ are negative as well. As discussed by Lederer and Takahashi[19] in the context of the Hubbard model, this choice of phase corresponds to the so-called "$s + id$" state.[20] Thus, in the absence of the holon kinetic energy, the effect of $H_{J'}$ is to cause each holon to bind a half flux quantum of statistical flux.

The holon number (topological charge[4(a)]) couples to the statistical gauge-field in the same way as electromagnetic charge couples to the electromagnetic gauge field, which follows from Eq. (10). To see this explicitly, imagine that there is a half integer flux quantum through the plaquette at the origin. We work in a singular gauge in which there is a Dirac sting running along the positive $y$ axis, which is to say $a(l) = \pi$ for the column of links which cross the string, and $a(l) = 0$ for all other links as shown in Fig. 1(a). Thus, the phase can be expressed as $e^{i\theta_c} = (-1)^{n_c}$ where $n_c$ is the number of dimers which cross the $y$ axis at positive $y$. As shown in Fig. 1(b), when a holon moves past the string, the number of dimers crossing the string changes by $\pm 1$. In other words, if a holon makes a closed orbit around the flux quantum, the wave function changes sign; this is just the Aharonov-Bohm phase corresponding to a half flux quantum.

Since all the bare particles in the theory are bosons, i.e., dimers which consist of a tightly bound electron pair, it is clear that the bare holon is a boson,[4] and this can be checked directly. Thus, in the usual way[1(a)] of two-dimensional systems, a holon-fluxoid bound state can be treated either as a boson, in which case the bosons interact via a long-ranged gauge force, or, more simply, as a weakly interacting fermion. This conclusion reproduces, in the context of the hard-core

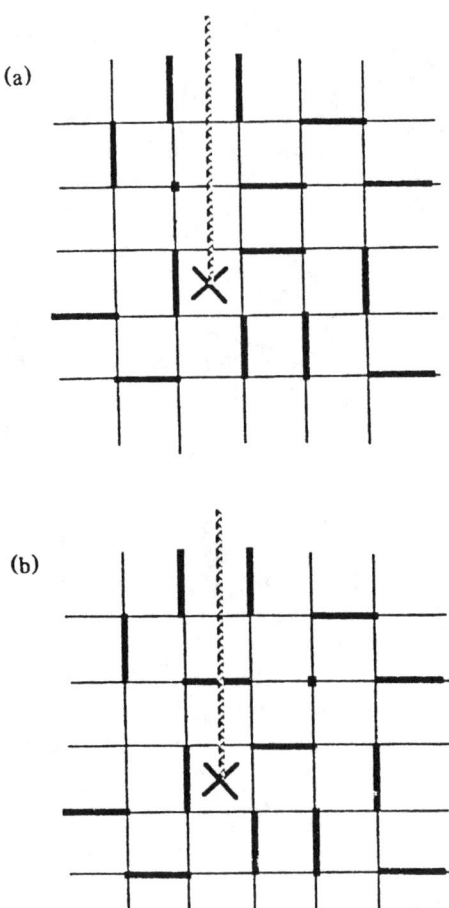

FIGURE 1 A vortex excitation at the origin (marked by an ×) with a Dirac sting lying along the positive y axis. Shown are representative dimer configurations (dimers are represented by bars and a holon by a dot) and information about the phases with which they enter the sum in Eq. (6). There is one half-quantum of statistical flux through each plaquette and an additional half flux quantum through the marked plaquette. This, $a(l) = \pi$ for all the links that are cut by the string. In (a) there are no dimers crossing the Dirac string, while after the holon crosses the string [in (b)] there is one.

dimer model, the earlier conclusions of Read and Chakraborty.[11] However, we see at once that the statistics of the low-energy state (quasiparticle) of the holon is determined by energetic considerations. If $J'$ were negative, then the bare holon would have lower energy by order of $|J'|$ than the holon-fluxoid pair; as a result the holon would be a boson. More to the point, in the presence of a nonzero

# Statistics of Holons in the Quantum Hard-Core Dimer Gas

holon kinetic energy, the holon will be a boson, even for positive $J'$, so long as $t \gg J'$. This follows from the fact that repeated holon hoping generates an effective coupling between the dimer states connected by the pinwheel operator as shown in Fig. 2; the holon kinetic energy is minimized if all dimer states which differ by the rearrangements of dimers in the vicinity of the holon enter with equal phases. (In the Appendix we estimate the critical value of the ratio of $t$ to $J'$ at which the unbinding transition occurs.) Again, this is consistent with the results of Lederer and Takahashi[19] who found in the context of the short-ranged RVB state in the Hubbard model that the holon kinetic energy is minimized if the superposition is locally "s-like" in the neighborhood of each holon. Since we expect physically that $t \gg J'$, the holons will generally be bosonic.

It is worth commenting briefly on the statistics of the spinons. Spinons are not present in the hard-core dimer model per se, since the model presupposes a gap in the spin excitation spectrum, and deals only with excitations at energies less than the spin gap.[5] Moreover, if we allow a finite density of spinon excitations, we are no longer guaranteed that the different valence-bond states are linearly independent; indeed the set of states consisting of nearest-neighbor valence-bonds plus unpaired spins is clearly overcomplete. However, at dilute spinon density, we can treat a spinon as a holon with an electron bound to it. Clearly, such a particle has electromagnetic charge 0 and spin $\frac{1}{2}$. It also carries the same topol-

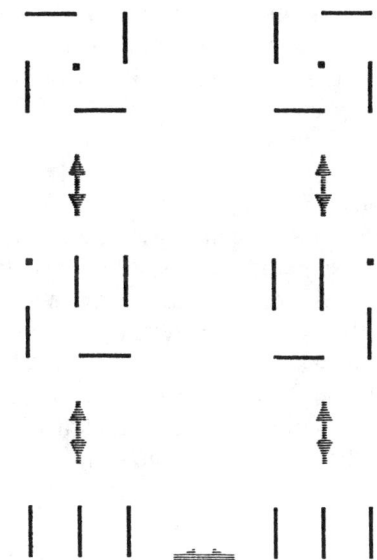

FIGURE 2 An effective pinwheel operation can be generated by repeated action of the holon kinetic energy.

ogical charge as the holon, so it couples in an identical fashion to the statistical gauge field. Thus, a bare spinon is a fermion while a spinon-fluxoid pair is a boson. Unlike the holon, however, the spinon kinetic energy is neither large compared with $J$ and $J'$, nor is it clear whether it favors local $s$-like or $s + id$-like symmetry. It seems natural, as suggested in Refs. 10 and 11, that the spinon will bind a fluxoid and hence will be bosonic, but this presumably depends on the degree of frustration in the model. At any rate, since whenever the short-ranged RVB analysis is applicable there must be a gap to the spinon states, we are never interested in the low-temperature properties of a dense spinon gas, and so the statistics of the spinons is not of primary importance.

## IV. SPIN, QUASISTATISTICS RELATION

In three dimensions, statistics are more robust than in two; we cannot simply change the statistics of the particles by adding a pure gauge field to the system as was the case in two dimensions. This suggests that the statistics can be determined on more fundamental ground. The spin-statistics theorem of relativistic quantum mechanics is such a connection. Unfortunately, its proof relies on Lorentz invariance. In the class of problems we are interested in, the spin is an internal quantum number of the bare particles of the problem, the electrons; it is not necessarily connected to any of the space-time symmetries of the system. The question we are interested in answering is; if the low-energy excitations of a collection of electrons have a quasiparticle description, are these quasiparticles required to have a particular spin (quasiparticle) statistics relation? In general, the answer must necessarily be that they do not. However, in three dimensions in particular, it *is* often the case that there is a connection between the spin and spatial rotational symmetries of the wave function. It is easy to see that if a two-quasiparticle wave function has any axis of symmetry such that it is invariant under a 180° rotation of both spin and space, then the quasiparticles must be bosons if they have integer spin and fermions if they have half-integer spin. This follows immediately from the fact that such a transformation interchanges the two quasiparticles. A rotation of 180° of two half integer spins produces a minus sign which, for an invariant wave function, must be canceled by a factor of $-1$ per exchange. Similarly, for integer spin, the exchange factor must be $+1$. Based on this argument, we feel that it is likely (though not proven) that in higher dimensions, the holons will be bosons and the spinons will be fermions.

## V. FLUX QUANTIZATION

An issue of general importance is the issue of flux quantization in systems with quasiparticles with fractional charge and/or statistics. It was argued in Ref. 21

that, quite generally, charge fractionalization will not produce any effect on the Aharonov-Bohm periodicity of observable quantities in a non-simply = connected system, since the quasiparticle wave functions will be multivalued in just such a way as to leave the flux quantum unchanged. The issues of flux quantization and the effects of statistical transmutation are somewhat more subtle since they address issues of macroscopic phase coherence. Arguments similar in spirit to those in Ref. 21 were presented in Refs. 4(b) and 22 to demonstrate that in the present model, the reverse charge-statistics relations of the holons will not alter the fact that flux quantization will occur in units of $hc/2e$. Indeed, it can *almost* be argued on general grounds[23] that if the only bare charged particles are fermions with charge $e$, then flux quantization will always occur in units of $hc/2ne$ where $n$ is an integer.

On the other hand, Wilzcek[1(a)] has shown that in two dimensions there is no way to distinguish a charge $e$ boson from a charge $e$ fermion bound to a half flux quantum. Yet there is no doubt that at zero temperature, a system of charge $e$ bosons can condense into a superfluid state with flux quantum $hc/e$.

The resolution of this seeming paradox can be understood along lines first suggested by Wen.[7] It derives from the existence of vortex excitations which carry charge zero and half a flux quantum. Consider the purely magnetic model in which $t$ is set equal to zero. We define the one vortex state to be the state in which an additional half flux quantum of statistical flux is threaded through an arbitrary plaquette, e.g., the plaquette at the origin as is Fig. 1(a). This state is orthogonal to the ground state,[11] and has an excitation energy of order $J$. Note that there is no magnetic energy associated with the Dirac string since so long as the holon positions are held fixed, any process which rearranges the dimers will change the number of dimers crossing the string by an integer multiple of 2, except only those which involve a net circulation of dimers about the vortex. It is the existence of these vortices which is responsible for the fact that flux quantization is in units of $hc/2e$.

To see this, consider the cylindrical geometry shown in Fig. 3. We consider the periodicity of the ground-state energy $E(\Phi)$ of a condensate of change-$e$ bosons which live on the surface of the cylinder as function of the electromagnetic flux $\Phi$ through the cylinder. Clearly, $E(\Phi)$ is periodic with period $hc/e$. However, there are two possible branches of $E(\Phi)$ as shown respectively in Figs. 3(a) and 3(b). If there is no statistical flux through the cylinder, then the energy is minimized when there is an integer number of electromagnetic flux quanta through the cylinder; if there is a half quantum of statistical flux through the cylinder (i.e., if there is a Dirac string that bisects the cylinder) then the energy is minimized when there is a half odd-integer number of electromagnetic flux quanta through the cylinder, as shown in Fig. 4. This follows immediately from the fact that electromagnetic and statistical flux couple to the holon current in identical fashion. The two branches of $E(\Phi)$ are identical other than the offset of their zeros, since there is no energy associated with a Dirac strong which has no ends. Were there no vortices, the system with half a statistical flux quantum through the cylinder could never mix with the fluxless system, and the only observable

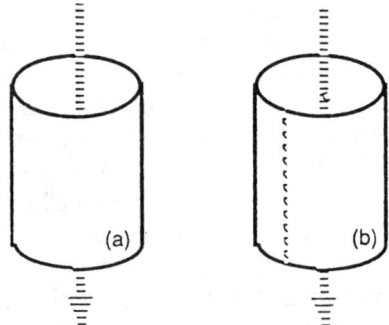

FIGURE 3 Geometry for the discussion of flux quantization. Here, the electromagnetic flux through the center of the cylinder (indicated by the arrow) is varied. In (a) there is no net statistical flux through the cylinder, while in (b) there is half a flux quantum of statistical flux as indicted by the presence of a Dirac string.

periodicity would be in units of $hc/e$. However, the presence of vortices with finite energy allows the system to tunnel from one sector to the other by creating a vortex-antivortex pair and pulling them off either edge of the cylinder. Thus, there is level mixing, and the true grounds-state energy has periodicity $hc/2e$, as shown in the dashed line in Fig. 4.

In the model Wilczek considered, there are no independent dynamical degrees of freedom corresponding to the vortices, which is equivalent in our model to setting the creation energy of a vortex to infinity. In that limit, but only in that limit, the superconducting flux quantum would be $hc/e$. This resolves the apparent paradox. However, if the vortex creation energy is large but not infinite, the mixing caused by the vortices could be small, and it might be possible to do a "fast" experiment on small superconducting rings and observe flux quantization in units of $hc/e$.

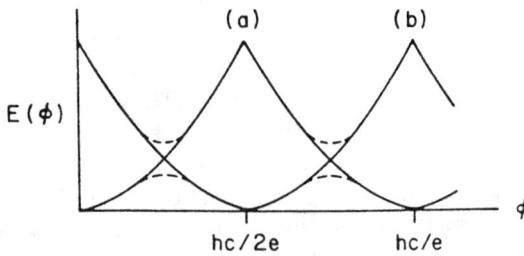

FIGURE 4 Ground-state energy as a function of the electromagnetic flux through the cylinder in Fig. 3. Curve a is the branch corresponding to no statistical flux through the cylinder, curve b to half a flux quantum. The dashed curve includes the effect of mixing between the first two curves.

## ACKNOWLEDGMENTS

I would like to acknowledge extremely useful discussions with E. Abrahmas, S. Coleman, E. Fradkin, F. D. M. Haldane, D. S. Rokhsar, S. Shenker, and most expecially with N. Read. This work was largely done at the Aspen Institute for Physics and was supported in part by National Science Foundation (NSF) Grant No. DMR-87-06250.

## APPENDIX: HOLON-FLUXOID INTERACTION

Here we wish to estimate $X$, the critical ratio of $J'/t$ such that for $J'/t > X$ the holon will bind a half flux quantum while for $J'/t < X$, the holon will be free. To do this, we compute the eigenstates and eigenvalues of the dimer Hamiltonian on the nine-site system pictured in Fig. 2, and we assume free boundary conditions. (Only a subset of the possible dimer configurations are pictured in the figure; in total there are 18 dimer states for free boundary conditions.) Of the 18 eigenstates of this Hamiltonian, only four have nonvanishing amplitude for the two dimer configurations with the holon on the central site. (For example, the configuration in the upper left-hand corner of the figure.) We will consider only these four states.

Of these four states, one of them is antisymmetric under rotation by $\pi/2$. This state is the lowest-energy state of the four for the pure magnetic Hamiltonian; its energy is $E_{\text{anti}} = -J'$, independent of the other interactions. The other three states are symmetric under rotation. Their energies are obtained from the solution of the following cubic equation:

$$(J' - E)(2V + J - E)(V - E) - J'J^2 - 4t^2(2V + J - E) = 0.$$

The lowest-energy solution of this equation is the symmetric ground state energy, $E_{\text{sym}}$.

The antisymmetric state is the state in which there is an extra half flux quantum associated with the holon, so the unbinding transition occurs when $E_{\text{sym}} = E_{\text{anti}}$. Since the solution of a cubic equation is awkward, and our estimate is crude at best, we choose to evaluate $E_{\text{sym}}$ for $V = J = 0$, so

$$E_{\text{sym}} \approx \frac{J'}{2} - \left[\left(\frac{J'}{2}\right)^2 + (2t)^2\right]^{1/2}.$$

In this approximation the critical value of $J'/t = \sqrt{2}/2$.

## REFERENCES

1. (a) R. Jackiw and C. Rebbi, *Phys. Rev. D* **13**, 3998 (1975); (b) F. Wilzcek, *Phys. Rev. Lett.* **49**, 957 (1982); (c) D. P. Arovas, F. Wilzcek, J. R. Schrieffer, and A. Zee, *Nucl. Phys.* **B251**, 117 (1985); (d) R. Jackiw, *Helv. Phys. Acta* **59**, 835 (1986); (e) G. Semenoff, *Phys. Rev. Lett.* **61**, 517 (1988).
2. R. B. Laughlin, *Phys. Rev. Lett.* **50**, 1395 (1983); B. I. Halperin, *ibid.* **52**, 1583 (1984); F. D. M. Haldane, *ibid.* **51**, 605 (1983); D. P. Arovas, F. Wilcek, and J. R. Schrieffer, *ibid.* **53**, 722 (1984).
3. P. W. Anderson, *Science* **235**, 1196 (1987); G. Baskaran, Z. Zhou, and P. W. Anderson, *Solid State Commun.* **63**, 973 (1987); S. Coppersmith and P. W. Anderson (unpublished); Z. Zhou and P. W. Anderson, *Phys. Rev. B* **37**, 627 (1987).
4. (a) S. Kovelson, D. S. Rokhsar, and J. P. Sethna, *Phys. Rev. B* **35**, 8865 (1987); (b) *Europhys. Lett.* **6**, 353 (1988); (c) S. Kivelson, *Phys. Rev. B* **36**, 7237 (1987).
5. D. S. Rokhsar and S. Sivelson, *Phys. Rev. Lett.* **61**, 2376 (1988); S. Kivelson and D. S. Rokhsar, *Physica C* **153**, 531 (1988); S. Kovelson and D. S. Rokhsar, *Phys. Rev. Lett.* **61**, 2630 (1988).
6. M. J. Rice and Y. R. Wang, *Phys. Rev. B* **37**, 5893 (1988); and (unpublished).
7. X-G. Wen (unpublished).
8. I. E. Dzyaloshinskii, A. M. Polyakov, and P. B. Weigman, *Phys. Lett. A* **137**, 112 (1988); P. B. Wiegman, *Phys. Rev. Lett.* **60**, 821 (1988).
9. V. Kalmeyer and R. B. Laughlin, *Phys. Rev. Lett.* **59**, 2095 (1987); R. B. Laughlin, *ibid.* **60**, 2677 (1988); E. J. Mele (unpublished).
10. F. D. M. Haldane and H. Levine (unpublished).
11. N. Read and B. Chakraborty (unpublished).
12. D. S. Rokhsar (unpublished).
13. W. Marshall, *Proc. R. Soc. London, Ser. A* **232**, 48 (1955).
14. J. T. Chayes, L. Chayes, and S. Kovelson, *Commun. Math. Phys.* (unpublished).
15. D. J. Klein, *J. Phys. A* **15**, 661 (1982).
16. P-O. Lowdin, *J. Chem. Phys.* **18**, 365 (1950).
17. G. Baskaran and P. W. Anderson, *Phys. Rev. B* **37**, 580 (1988).
18. E. Fradkin and S. A. Kivelson (unpublished).
19. P. Lederer and Y. Takahashi (unpublished).
20. G. Kotliar, *Phys. Rev. B* **37**, 3664 (1988).
21. S. Kivelson and M. Rocek, *Phys. Lett.* **156B**, 85 (1985).
22. D. Thouless, *Phys. Rev. B* **36**, 7187 (1987).
23. C. N. Yang, *Rev. Mod. Phys.* **34**, 694 (1962).

**Gabriel Kotliar**
Department of Physics
Massachussetts Institute of Technology
Cambridge, MA 02139

and

**Sandro Sorella**
International School For Advanced Studies
Strada Costiera 11
34014 Trieste - Italy

# Conductivity and Tunnelling Density of States Exponents at the Metal Insulator Transition with Strong Spin Orbit Scattering

## INTRODUCTION

Our understanding of the metal insulator transition in interacting disordered Fermi systems has increased enormously recently.[1] The mapping of the metal-insulator transition problem in the presence of randomness and electron-electron interactions into a generalized nonlinear σ-model[2] allowed the identification of the scaling variables and provided an efficient tool for generating a perturbation theory in powers of the dimensionless resistance $t \equiv \frac{e^2}{2\pi^2\hbar} R$. The results of the nonlinear σ-model have been shown to be consistent with diagrammatic perturbation theory satisfying all the Ward identities deriving from the global gauge invariance of the model.[3] The scaling variables relevant at the metal-insulator transition have been interpreted in terms of quasi-particle parameters in a Landau-like framework.[4]

There are several universality classes describing different physical scenarios for a metal-insulator transition.[5] The different universality classes are determined by the conservation laws, the range of the electron-electron interactions and by the symmetries, which in turn depend on whether the system has magnetic impurities, spin orbit scattering, or is in the presence of strong or weak magnetic fields.

In this paper we reconsider the metal-insulator transition in the presence of spin orbit scattering and electron-electron interaction. We provide a detailed derivation of the renormalization group ($RG$) equations first derived by Castellani et al.[6] We use a parametrization which makes the calculations more similar to the field theory calculations in more conventional $\sigma$ models.[8] This parametrization allows us to calculate the singular corrections to the one particle density of states and its critical index to one loop order.[7]

The only detailed measurements of the tunnelling density of state close to the metal insulator transition have been carried out in systems having strong spin orbit scattering[9,10] and we compare the results of the $\epsilon$ expansion with the experiments.

The content of the paper is the following: in Section 1 we outline the derivation of the renormalization group equations in the presence of spin orbit scattering. For pedagogical reasons we present a detailed derivation of the perturbation theory of the nonlinear $\sigma$ model describing the diffusion modes of interacting electrons. Section 2 contains the evaluation of the critical index $\nu$ to order $O(\epsilon^{3/2})$. In Section 3 we discuss the one particle density of states corrections. We conclude in Section 4 with a comparison of the theoretical predictions and the results of tunneling experiments.

## I. THE PERTURBATION THEORY AND THE RENORMALIZATION GROUP EQUATIONS

The starting point is the field theoretical formulation of the interacting disordered Fermi system due to Finkel'stein.[2] The partition function of the system is given by a functional integral over a matrix field $Q$ to be specified below.

$$Z = \int DQ e^{-(L_d[Q] + L_{ee}[Q])} \tag{1}$$

The Lagrangian

$$L_d[Q] = \frac{1}{8t} \int dr tr \nabla Q \nabla Q - \frac{z}{2t} \int tr \hat{\epsilon} Q$$

$$L_{ee} = -\frac{1}{2t} \int dr \left[ \Gamma_s \sum_{a=0,3} Q^a \gamma_d Q^a (-1)^a + \frac{\Gamma_c}{2} \sum_{a=1,2} Q^a \gamma_c Q^a \right] \tag{2}$$

contains four parameters: $t$ is the inverse diffusion coefficient, $z$ is related to the renormalization of the specific heat,[5] $\Gamma_s$ is the singlet scattering amplitude, $\Gamma_c$ is the scattering amplitude in the Cooper channel. These scaling parameters have been recently reinterpreted in terms of a Fermi liquid framework for a disordered system.[5] $\hat{\epsilon}$ is a matrix in the frequency space, replica space and is proportional to the identity quaternion $\hat{\epsilon}_{nm} = \delta_{nm}\epsilon_n \delta_{ij}\tau_0$, $\epsilon_n$ is a Matsubara frequency, $(2n +$

## Conductivity and Tunnelling Density of States Exponents

$1)\pi T$. $Q$ is a matrix in frequency space, and replica indices. Each matrix element is a quaternion of the form

$$Q_{ni,mj} = Q^0_{ni,mj}\tau_0 + iQ^1_{ni,mj}\tau_1 + iQ^2_{ni,mj}\tau_2 + Q^3_{ni,mj}\tau_3 \tag{3}$$

with

$$\tau_0 = \begin{pmatrix} 1 & 0 \\ 0 & 1 \end{pmatrix} \quad \tau_1 = \begin{pmatrix} 0 & -i \\ -i & 0 \end{pmatrix} \quad \tau_2 = \begin{pmatrix} 0 & -1 \\ 1 & 0 \end{pmatrix} \quad \tau_3 = \begin{pmatrix} -i & 0 \\ 0 & i \end{pmatrix}$$

matrices obeying the quaternion algebra, $\tau_0$ being the identity and the remaining matrices satisfying $\tau_i\tau_j = \epsilon_{ijk}\tau_k$. The operators $\gamma_d$, $\gamma_c$ act on matrix indices and are given by

$$Q\gamma_d Q = 2\pi T \sum_{n_1 n_2 n_3 n_4 i} Q_{n_1 i n_2 i} q_{n_3 i n_4 i} \delta_{n_1+n_3, n_2+n_4}$$
$$Q\gamma_c Q = 2\pi T \sum_{n_1 n_2 n_3 n_4 i} Q_{n_1 i n_2 i} Q_{n_3 i n_4 i} \delta_{n_2+n_3, n_2+n_4} \tag{4}$$

The functional integral in (1) is restricted to hermitian matrices obeying

$$Q^+ = Q \quad Q^2 = I \quad \text{tr } Q = 0 \tag{5}$$

The first condition in (5) implies a $Q^a_{ni,mj} = Q^a_{mj,ni}$, $a = 0, 1, 2$ and $Q^3_{ni,mj} = -Q^3_{mj,ni}$, while because of the charge conjugation symmetry $Q^i$ are real matrices.[6] These matrices are represented as a unitary transformation $Q = U^+\Lambda U$ of the saddle point matrix $\Lambda_{ni,mj} = \delta_{nm} \text{sign}(n)\delta_{ij}$ and describe the soft modes of the problem. They can be parametrized in terms of *unconstrained* quaternion[12] matrices $V$ connecting positive and negative frequencies only.

$$Q = \begin{bmatrix} \sqrt{I - VV^+} & V \\ V^+ & -\sqrt{I - V^+V} \end{bmatrix} \tag{6a}$$

$$V_{ni,mj} = V^0_{ni,mj}\tau_0 + i\tau_1 V^1_{ni,mj} + i\tau_2 V^2_{ni,mj} + \tau_3 V^3_{ni,mj} \tag{6b}$$

Using this parametrization it is easy to verify that the constraints (5) are fully satisfied, while the quaternion structure (6b) is preserved in (3) because the productg (and any algebraic operation as in 6a) of two quaternions is again a quaternion of the same type (6b). Introducing the representation 6 in Eqs. 1–2 reduces the problem to functional integration over four real matrix fields $V^a_{ni,mj}$, $a = 0, 1, 2, 3$ which can be treated by standard perturbation theory. The parameterization of Eqs. (6) has the additional advantage of allowing the calculation of the renormalization of the one particle density of states which has not been calculated before in the spin orbit case. To carry out the perturbation theory we separate the quadratic part

$$L_d = \frac{1}{2t} \sum_{i=0}^{3} \int \nabla V^i_{nm} \nabla V^i_{nm} + z(\epsilon_n - \epsilon_m) V^i_{nm} V^i_{nm}$$

$$L_{ee} = \frac{2\pi t \Gamma_s}{t} \sum_{n_1 n_2 n_3 n_4} \delta_{n_1+n_3,n_2+n_4} \sum_{i=0,3} V^i_{n_1 n_2} V^i_{n_4 n_3} \quad (7)$$

$$- \frac{2\pi t \Gamma_c}{t} \sum_{n_1 n_2 n_3 n_4} \delta_{n_1+n_2,n_3+n_4} \sum_{i=1,2} V^i_{n_1 n_2} V^i_{n_4 n_3}$$

which defines the propagators of the theory.

$$\langle V^i_{n_1 n_2} V^i_{n_4 n_3} \rangle = tL(q,\omega) \delta_{n_1 n_2} \delta_{n_2 n_2} + 4\pi T t \Gamma_s(q,\omega) L^2(q,\omega) \delta_{n_1+n_3,n_2+n_4}$$

$$i = 0, 3 \quad (8)$$

$$\langle V^i_{n_1 n_2} V^i_{n_4 n_3} \rangle = tL(q,\omega) \delta_{n_1 n_4} \delta_{n_2 n_3} - 4\pi T t \Gamma_c(q,\omega) L^2(q,\omega) \delta_{n_1+n_2,n_3+n_4}$$

$$i = 1, 2 \quad (9)$$

with

$$L(q,\omega) = \frac{1}{(q^2 + z\omega)} \quad \Gamma_s(q,\omega) = \frac{\Gamma_s(q^2 + z\omega)}{q^2 + (z - 2\Gamma_s)\omega}$$

and

$$\Gamma_c(q,\omega) = \frac{\Gamma_c}{1 + \frac{\Gamma_c}{z} \ln \frac{\Lambda}{q^2 + z\omega}} \quad \text{with} \quad \epsilon_{n_4} \epsilon_{n_1} > 0 \quad \epsilon_{n_3} \epsilon_{n_2} < 0 \quad (10)$$

$\omega = \epsilon_{n_1} - \epsilon_{n_2}$, $\Lambda$ is an ultraviolet cutoff. The part of the propagator which is diagonal in the energy indices is denoted by a wavy line in Fig. 1a while the non-diagonal part of the propagator is represented in Fig. 1b, the dot denotes an amplitude $\Gamma_s$ or $\Gamma_c$.

The interaction vertices between the diffusion modes are obtained by inserting the parameterization (6) for the order parameter in the Lagrangian of Eq. (1) and collecting the terms which are cubic and quartic in the field $V$ which are needed for a one loop calculation. The interaction vertices are shown in Fig. 1. Diagram 1c is generated by expansion of $L_d$ while the terms generated by expansion of $L_{ee}$ are represented by diagram 1b–1h. The notation here follows

# Conductivity and Tunnelling Density of States Exponents 129

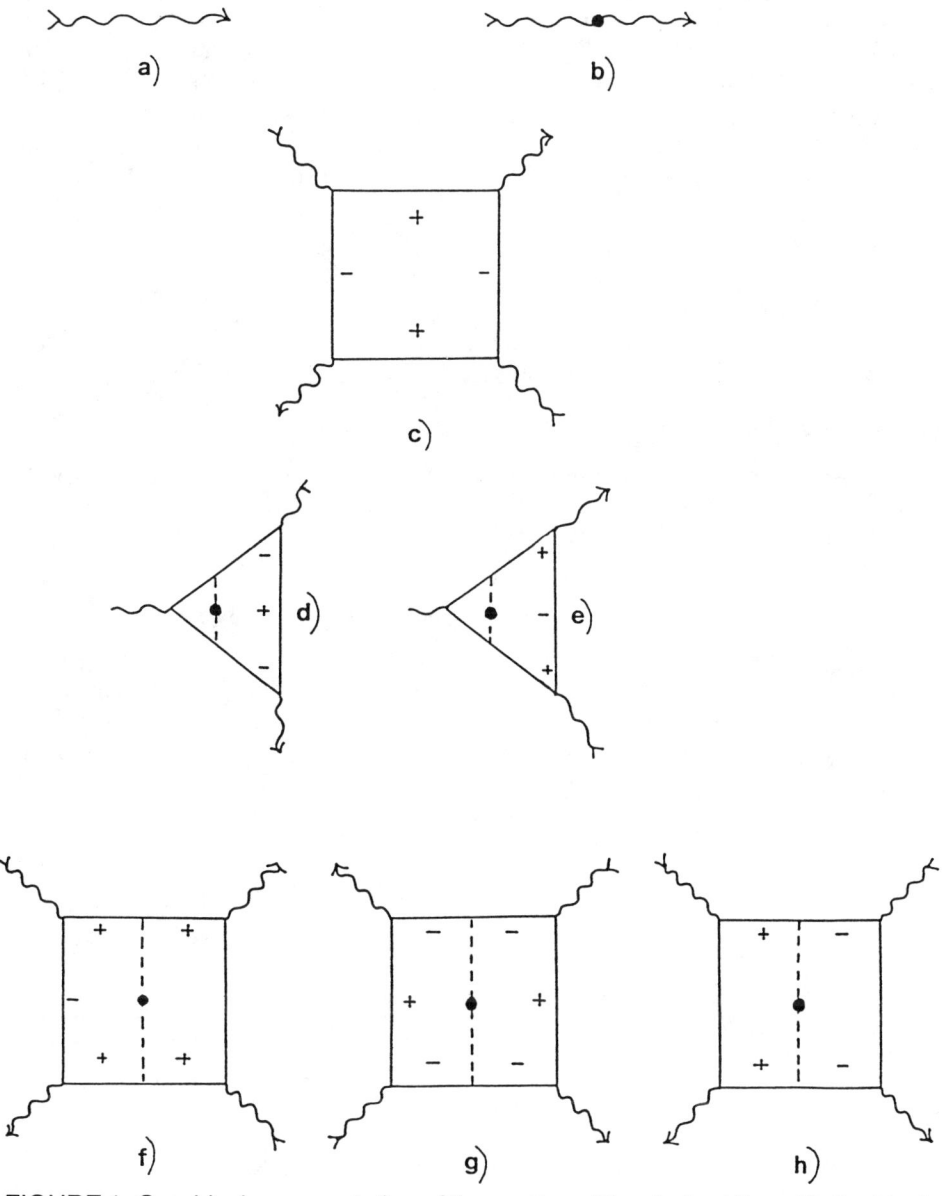

FIGURE 1 Graphical representation of the vertices. The dashed line with the black dot stands for a dynamical amplitude with the related energy and spin conservation laws. The incoming and outcoming arrows indicate formally the energy spins of the sides in a given vertex. For simplicity these signs are explicitly drawn.

Ref. (11), the wavy lines represent the quaternion field $V$. The $\Gamma_s$ 4-legged vertices have $V^0$ and $V^3$, $V^2$ and $V^1$ or $V^i$ and $V^i$ $i = 0\ 1\ 2\ 3$ lines on opposite sides of the amplitude dashed line. While the $\Gamma_c$ 4-legged vertices have $V^3$ and $V^1$, $V^2$ and $V^0$, $V^3$ and $V^2$ fields attached to the right or the left of the amplitude dashed line.

The one particle density of states is evaluated[2] from the expectation value of the matrix $Q_{nn}^0$. Using Eq. (6) we find to one loop order:

$$\langle Q_{nn}^0 \rangle = 1 - \frac{1}{2} \sum_i \langle (V^i V^{i+})_{nn} \rangle \tag{11}$$

The corresponding diagrams are shown in Fig. (2). The line denotes an insertion of $\sum_{mi} V^i_{nm}(q) V^i_{nm}(-q)$ the $i = 0, 3$ propagators are contracted with a singlet amplitude while the $i = 1, 2$ propagators are contracted with a Cooper ladder. Requiring $\xi^{-1}\langle Q_{nn}^0\rangle$ to be nonsingular we find for, $\xi = \dfrac{N}{N_o}$, which is the ratio of the bare to the renormalized density of states:

$$\xi = 1 - \frac{t}{2} \int \frac{d^d q}{\pi^{d/2}} \int d\omega [2\Gamma_s(\omega, q) - 2\Gamma_c(\omega, q)] L^2(q, \omega) \tag{12}$$

We work to lowest order to $\Gamma_c$ and perform the frequency integrations which are, to this order, independent of the ultraviolet cutoff. The remaining momentum integral is infrared divergent and is taken between the upper cutoff $\Lambda$ and the lower momentum cutoff $\lambda\Lambda$ ($\lambda\Lambda < q < \Lambda$).

$$\xi = 1 + \frac{t\Lambda_c}{z} \ln\frac{1}{\lambda^2} - \frac{t}{\epsilon}\left[\left(1 - \frac{2\Gamma_s}{z}\right)^{\epsilon/2} - 1\right]\ln\frac{1}{\lambda^2} \tag{13}$$

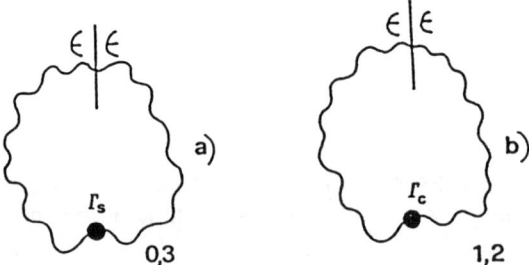

FIGURE 2 DOS diagrams at one loop order. The slashes mean that either the upper or the lower frequency of the two separated propagators has to be fixed. The numbers written near the dynamical amplitudes stands for the possible quaternion components to be considered in the contractions.

contracting the vertices in Fig. 1b with the propagators in Fig. 1a we obtain the one loop renormalization of the propagator (see Fig. 2).

The diagrams which are diagonal in the energy indices, termed self energy diagrams, are shown in Fig. 3. The inverse propagator of the theory which is diagonal in the energy indices is expected to scale as $L^{-1}(\omega = 0) = \xi^{-2}[Dk^2]$. Evaluating the diagrams in Fig. 3 we find

$$\delta(\xi^{-2}Dk^2) = \frac{k^2}{2} \int \frac{dq^d}{\pi^{d/2}} L(q, 0)$$

$$- \int d\omega \frac{dq^d}{\pi^{d/2}} [L(q + k, \omega) - (k^2 L^2(q, \omega) + L(q, \omega))][\Gamma_s(q, \omega) - \Gamma_c(q, \omega)] \quad (14)$$

combining Eqs. (13) and (14) one finds the correction to the diffusion constant

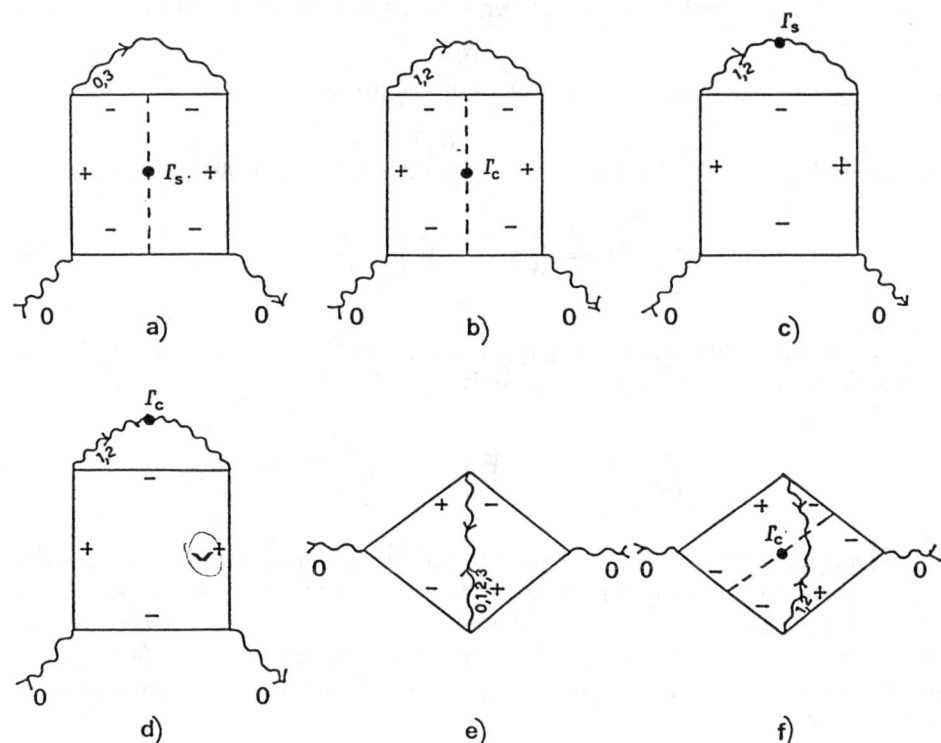

FIGURE 3 One loop order self energy diagrams. Here the notation are the same as in Fig. 2. In such diagrams the upper and lower energy indices are independently conserved in the external legs. The external quaternion indices are fixed to the identity component.

$$\delta D = \left\{ \frac{1}{2} - \left[ 1 + \frac{1 - \frac{2\Gamma_s}{z}}{\frac{2\Gamma_s}{z}} \ln\left(1 - \frac{2\Gamma_s}{z}\right) \right] - \frac{\nu\Gamma_c}{z} \right\} \ln \lambda^{-2} \qquad (15)$$

where $\lambda$ is again the ratio of the upper and lower momentum cutoffs in the integral.

We now consider the correction to the propagators $\langle V_i V_i \rangle$ which are off diagonal in the energy indices, for $i = 0, 3$ and $i = 1, 2$ those are the corrections to the singlet and Cooper amplitude respectively. The diagrams which renormalize the Cooper amplitude are shown in Fig. 4. Diagrams (c), (d) and (e) add up to

$$\delta(\xi^{-2}\Gamma_c)_{c-e} = -2t \int dg \Gamma_s(\omega, q) L^2(q, \omega) \qquad (16)$$

This term is the renormalization of the density of states due to the singlet part of the interaction. Notice that there are no $\Gamma_c^2$ diagrams and the Cooper contribution to the one particle density of states is not canceled. (a) and (b) add up to

$$\delta(\xi^{-2}\Gamma_c)_{a-b} = t \int d^d q \Gamma_s(q, 0) L(q, 0) = t\Gamma_s \ln \lambda^{-2} \qquad (17)$$

combining Eqs. 16–17 with Eq. 12 we find $\delta\Gamma_c = \delta(\xi^{-2}\Gamma_c) + 2\delta(\xi)\Gamma_c$, that is

$$\delta\Gamma_c = t\left(2\frac{\Gamma_c^2}{z} + \Gamma_s\right) \ln \frac{1}{\lambda^2} \qquad (18)$$

To that one has to add the usual logarithmic contribution coming from the Cooper channel summation, see diagram (f), arriving at

$$\delta\Gamma_c = t\left(2\frac{\Gamma_c^2}{z} + \Gamma_s\right) \ln \frac{1}{\lambda^2} - \frac{\Gamma_c^2}{z} \ln \frac{1}{\lambda^2} \qquad (19)$$

Finally, we turn to the corrections to the singlet amplitude. In this case the diagrams coming from the contraction of the three point vertices (5a, 5b and 5c) exactly cancel the DOS contribution (both the singlet and the Cooper corrections) to the renormalized coupling $\Gamma_s$. The remaining diagrams (5d and 5e) directly provide a correction to the singlet amplitude after the mentioned cancellation of the DOS contribution:

$$\delta\Gamma_s = -\frac{t}{2} \int \frac{dq^d}{\pi^{d/2}} L(q^2, 0)(\nu\Gamma_s - \nu\Gamma_c) \qquad (25)$$

We also check that the coupling $z$ renormalizes as $2\Gamma_s$. Combining Eqs. (15),

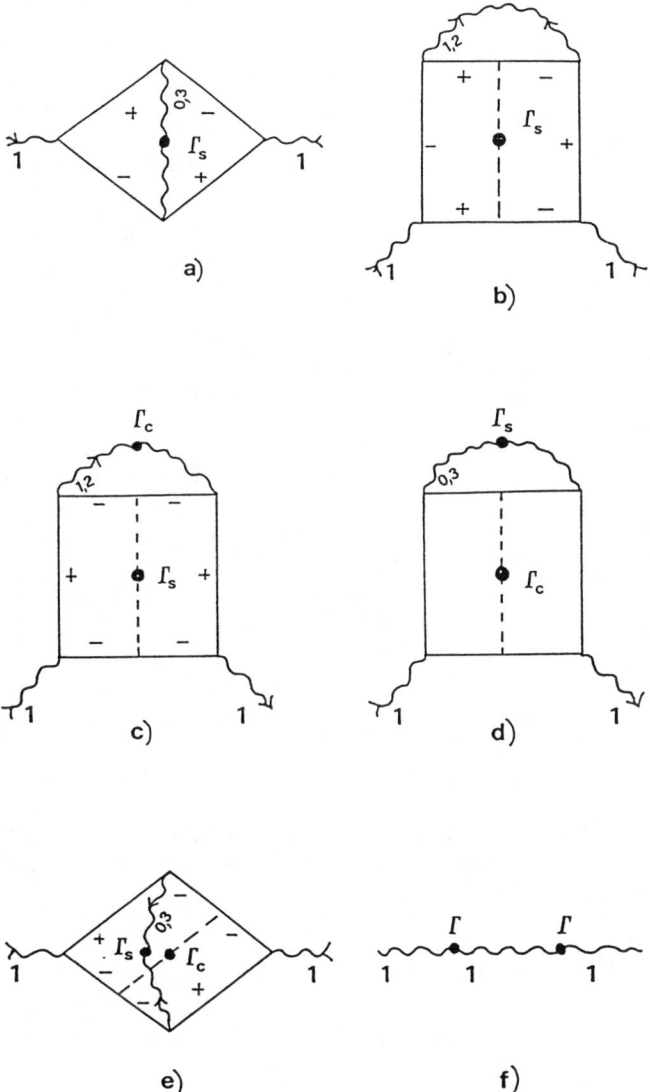

FIGURE 4 One loop order diagrams for the Cooper amplitude renormalization. Here we follow the same notation as in Fig. 2 while the external quaternion indices is fixed to the component $\tau_1$. The energy conservation of the external legs must not be diagonal in the energy indices (this would be a self energy renormalization).

(19) and (25) one finds the R.G. equation for the couplings $\Gamma_s$, $D$, $\Gamma_c$ of Castellani et al.[3] by taking the derivative of the recursion relations with respect to the logarithm of the inverse rescaling parameter $\ln(\lambda^{-1})$

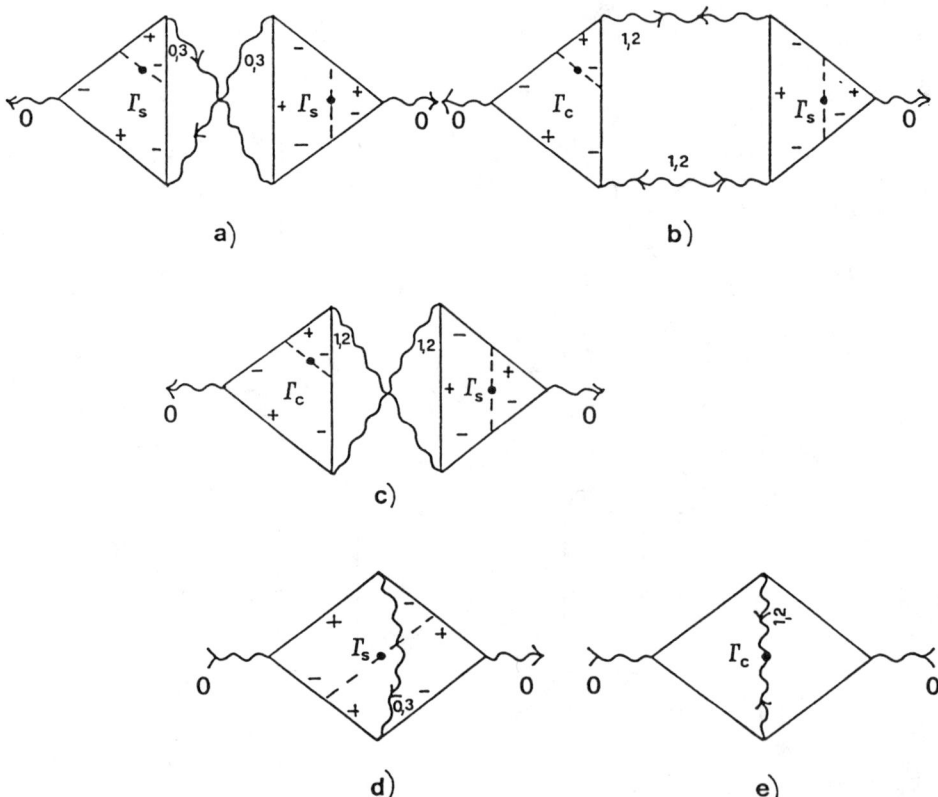

FIGURE 5 One loop order diagrams for the singlet amplitude renormalization. The notation are the same as in Fig. 2. The external quaternion indices are fixed to the identity component.

$$\dot{D} = \left\{ 1 - \left[ 2 + \frac{1 - \frac{2\Gamma_s}{z}}{\frac{\Gamma_s}{z}} \ln\left(1 - \frac{2\Gamma_s}{z}\right) \right] - \frac{2\Gamma_c}{z} \right\}$$

$$\dot{\Gamma}_s = -t[\Gamma_s - \Gamma_c] \tag{27}$$

$$\dot{\Gamma}_c = t\left[ \frac{4\Gamma_c^2}{z} + 2\Gamma_s \right] - \frac{2\Gamma_c^2}{z} \tag{28}$$

## II. THE CRITICAL EXPONENTS

We consider here the long range case where $2\Gamma_s = z$, a condition which is preserved by the R.G. eqs.

We rewrite Eqs. (26)–(28) in terms of only two variables $t = \dfrac{1}{D}$ and $\gamma = \dfrac{\Gamma_c}{z}$. In $2 + \epsilon$ dimensions the equations for $t$ acquires an additional term from its bare dimensionality.

$$\dot{\gamma} = [-2\gamma^2 + t(1 + \gamma + 2\gamma^2)] = \beta_\gamma(\gamma, t) \qquad (29)$$

$$\dot{t} = -\epsilon t + t^2(1 - 2\gamma) = \beta_t(\gamma, t) \qquad (30)$$

Here we notice that these renormalization group equations which are valid to one loop order allows the determination of the critical indices to order $\epsilon^{3/2}$ since there is an unstable fixed point at

$$t^* = \epsilon(1 + \sqrt{2\epsilon}) \qquad \gamma^* = \sqrt{\frac{\epsilon}{2}} + \frac{3}{4}\epsilon \qquad (31)$$

At this fixed point $z$ scales to zero as $z(\lambda) \sim \gamma^{-x}$ as $\lambda \to 0$ with $x = -t^*(1 - 2\gamma^*) = -\epsilon + O(\epsilon^2)$; $x$ appears in the specific heat exponent $C_V \propto T^{\frac{-x}{d+x}}$ at the transition. Linearizing around the fixed point we find the critical index $\lambda^{-1} \propto \xi_c \propto (t - t^*)^{-\nu} \propto (n - n_c)^{-\nu}$

$$\frac{1}{\nu} = \epsilon\left[1 - \sqrt{\frac{\epsilon}{2}}\right] \qquad (32)$$

where $\xi_c$ is the correlation length and $n$ is the impurity concentration. Then the conductivity index is determined as

$$\sigma \sim (n - n_c)^s \qquad s = \epsilon\nu = 1 + \sqrt{\frac{\epsilon}{2}} \qquad (33)$$

## III. THE ONE PARTICLE DENSITY OF STATES

The density of states exponents are obtained following Ref. 2. We rewrite Eq. (13) as

$$\frac{\delta N}{N} = \left[\frac{t}{\epsilon}\left[\left(1 - \frac{2\Gamma_s}{z}\right)^{\epsilon/2} - 1\right] + \frac{t\Gamma_c}{z}\right]\ln\frac{1}{\lambda^2} \quad (34)$$

In the long range case $\frac{2\Gamma_s}{z} = 1$ and the logarithmic series has a coefficient of order $\frac{t^*}{\epsilon}$; as pointed out first by Finkel'stein.[2]

$$\frac{\delta N}{N} = -\left[\frac{t^*}{\epsilon} + o(\epsilon)\right]\ln\frac{1}{\lambda^2} \quad (35)$$

which at the critical point exponentiate to $N(\lambda) \propto \lambda^\theta$

$$\theta = 2 + 2\sqrt{2\epsilon} \quad (37)$$

The R.G. analysis suggests a scaling form for the one particle density of states[2]

$$N(T, \delta n) = \lambda^\theta N\left(\frac{T}{\lambda^{d+x}}, \delta n\lambda^{-\frac{1}{\nu}}\right) \quad d + x = 2 + o(\epsilon^2) \quad (39)$$

and setting $\lambda = (\delta n)^\nu$ we find the exponent which characterize the vanishing of the density of states at the Fermi surface as one approaches to metal insulator transition,

$$N(0) \propto \delta n^{\nu\theta} \quad with \quad \nu\theta = \frac{2}{\epsilon}\left(1 + \frac{3}{\sqrt{2}}\epsilon^{1/2}\right) \quad (40)$$

On the other hand from Eq. 39 $\lambda$ scales as $T^{\frac{1}{d+x}}$, from which we derive the vanishing of the one particle density of states at the transition $n = n_c$ as a function of temperature with an exponent $N(T) \propto T^\beta$:

$$\beta = \frac{\theta}{d + x} = 1 + \sqrt{2\epsilon} \quad (41)$$

## CONCLUSIONS

Tunneling experiments were performed in NbSi films[10] close to the metal insulator transition. Conductivity measurements in the same films are consistent with $\sigma = (n - n_c)^s$ with $s \simeq 1$. The tunneling characteristics show metallic behaviour, i.e. $N(E) = N(0)\left[1 + \sqrt{\dfrac{E}{\Delta}}\right]$, for energies $E$ less than a crossover energy scale $\Delta$, $\Delta$ is seen to scale roughly as the second power of $(n - n_c)$. This agrees well with the results of the $\epsilon$ expansion because such exponent should be given, using Eq. (39), by $\nu(d + x) \simeq 2$. $N(0)$ is seen to scale as $(n - n_c)$, giving $\nu\theta = 1$. This exponent is very far from the result of Eq. (40). The fact that the $\epsilon$ expansion truncated to lowest order does not provide an accurate estimate of the critical exponents is not surprising since even for magnetic systems the $2 + \epsilon$ expansion around the lower critical dimension is far less accurate than the expansion around the upper critical dimension. The presence of fractional powers in the $\epsilon$ expansion of the critical exponents makes the nature of the series even more difficult to interpret, and we cannot extrapolate to the physical dimension $d = 3$, $\epsilon = 1$ with reasonable accuracy.

It is hard to extract from the data (10) the value of the exponent $\beta$. The value $\beta = \dfrac{1}{3}$ quoted in Ref. (10) is inconsistent with the scaling law $(d + x)\beta = \theta$ (see Eq. 41) which we derived from Eq. 39 and first obtained in Ref. 2. If the experimental value of $\beta$ is indeed $\dfrac{1}{3}$ one would have to explain theoretically the violations of the simple scaling law. It is important to note that the ratio of the short range part of the charge compressibility (which is the compressibility of the electron gas and its neutralizing backround) to the specific heat, denoted by $\dfrac{z_1}{z}$ in Ref. 4 diverges at the transition. One could suspect violations of scaling if this physical quantity was represented by a dangerously irrelevant operators that entered the expression for the density of states, in which case the extraction of the exponent $\theta$ using Eq. (39) would be incorrect. However, we checked that the two loop order most divergent term (i.e. to second order in $t$) in the expansion of the density of states is consistent with the lowest order calculation giving a strong support to the procedure for extracting the critical index of the density of states originally suggested by Finkel'stein.[2]

Another possibility is the scaling as a function of temperature (which is calculated in the theory as in Eq. (41)) is different from the scaling of the density of states at criticality as a function of energy since a finite temperature and a finite distance from the fermi level correspond to moving away from criticality in different eigendirections of the renormalization group. This hypothesis would be confirmed if a careful measurement of the density of states at the Fermi surface at the metal-insulator transition as a function of temperature would give a critical index $\beta$ different from $\frac{1}{3}$. More theoretical and experimental work is needed to elucidate this issue.

We obtained for the first time a nontrivial correction to the index $s$ for the conductivity in the interacting case. It differs from the usual ones $\nu = \frac{1}{\epsilon}$, $\nu = \frac{1}{2\epsilon} - \frac{3}{4} + O(\epsilon)$ calculated in the noninteracting case.

Notice that to order $\epsilon^{3/2}$ the quasi particle diffusion constant[4] $\frac{D}{z}$ is unrenormalized. Since the equations for $z$ and $t$ are rather different, we don't expect this equality to be valid to all orders. A measurement of the thermal diffusion constant or a two loop calculation is therefore necessary to analyze the critical behaviour of the quasi-particle diffusion constant close to the metal insulator transition.

## ACKNOWLEDGMENTS

We are deeply in debt with Prof. C. Castellani for useful discussions. One of us (S.S.) acknowledge the hospitality of MIT in Cambridge. This work was supported by the National Science foundation under grant number DMR-8521377 and partly under the sponsorship of the Italian Ministry of Education.

## REFERENCES

1. For a review see P. A. Lee and T. V. Ramakrishnan, *Rev. Mod. Phys.* **57**, 287 (1985).
2. Finkel'stein, *Sov. Phys. JEPT*, **57**, 97 (1985); **59**, 212 (1984).
3. C. Castellani, C. Di Castro, P. A. Lee, M. Ma, S. Sorella e E. Tabet, *Phys. Rev. B* **33**, 9 (1985); *Phys. Rev. B* **24**, 6783 (1984).
4. C. Castellani, G. Kotliar and P. A. Lee, *Phys. Rev. Lett.* **59**, 323 (1987).
5. For a review see C. Castellani and C. Di Castro in *Localization and Metal Insulator Transition*, H. Firtzshe and D. Adler, Plenum (1985).
6. C. Castellani, C. Di Castro, G. Forgacs and S. Sorella, *Solid State Communications* **52**, 261 (1984).
7. G. Kotliar and S. Sorella, submitted to *Phys. Rev. B*.
8. S. Hikami, *Phys. Rev. B* **24**, 2671 (1961).
9. W. L. McMillan and J. Mochel, *Phys. Rev. Lett.* **46**, 556 (1981).
10. D. Bishop, E. Spencer and R. C. Dynes, *Solid State Electronics* **28**, 173 (1985).
11. M. Grilli and S. Sorella, *Nucl. Phys.* **B295FS21**, 422 (1988).
12. K. Efetov, A. Larkin and D. Khmelnitzky, *Zh. Eksp. Teor. Fiz.* **79**, 1120 (1980). (*Sov. Phys. JEPT* **52**, 568 (1980)).

**Daniel C. Mattis**
Department of Physics, University of Utah
Salt Lake City, UT 84112, USA

# Coulomb Potential in Layered Metals, Bloch States, Wannier Functions and High-$T_c$ Superconductivity

The Coulomb interaction between charge carriers is not necessarily repulsive in reciprocal space (witness the attractive Madelung potential.) We prove that $V(\mathbf{q}) \neq 4\pi e^2/q^2$ at finite $\mathbf{q}$, and show that $V(\mathbf{q})$ can be negative near the surface of the Brillouin Zone. After appropriate dielectric screening, this interaction is shown to be capable of promoting Cooper pairing and superconductivity. The effective pair-potential is computed from first principles, starting with scattering amplitudes of the two-body Coulomb interaction in the representation of Bloch states. The effects are most pronounced for holes in quasi-two dimensional bands—just as is found experimentally.

## INTRODUCTION

The $CuO_2$-based high-$T_c$ superconductors are generally believed to owe their spectacular properties to a pairing mechanism other than the phonon-exchange mechanism of Bardeen, Cooper and Schrieffer.[1] Much attention has been paid to the natural antiferromagnetism of planes of $CuO_2$ in the $Cu^{+2}$ (spin 1/2), $O^{-2}$ (spin zero) valence states, and to the possible connections between antiferromagnetism and this new form of superconductivity. However well motivated such theories may be, I would like to draw attention to a more fundamental force at the very root of any conceivable magnetism: the ubiquitous Coulomb interaction. Consider the following set of conditions which favor high-$T_c$ superconductivity *according to the present theory*:

I. Layered materials in which the band structure is predominantly two-dimensional are favored. Inter-layer spacings $d$ must be just right.

II. Carriers should be holes, rather than electrons.

III. Electrical resistance in the normal state ($T > T_c$) should be anisotropic, with planar resistivity far lower than expected for strong-coupled superconductors at an equivalent density **n** of carriers.[2]

Calculations on the basis of this mechanism yield a distribution of gaps, $P(\Delta)$. The critical temperature $T_c$ increases with **n**, the density of hole charge carriers, up to a maximum value, beyond which it drops once again. In its crudest embodiment, the formula for $T_c$ takes on a familiar[1] form, illustrated in Fig. 1. With $\lambda$ required to be positive, it is:

$$kT_c = 1.13 \, \epsilon_F \exp -(1/\lambda) \tag{1}$$

with $\epsilon_F \approx \hbar^2 k_F^2/2m^*$. In two dimensions, $k_F^2 \sim$ **n** and (1) reduces to $T_c \alpha \mathbf{n}/m^*$ over the range of concentration where $\lambda$ is independent of $k_F$. *Such a proportionality has already been established experimentally.*[3]

In these pages we shall describe several other theoretical prescriptions as well, which are in equally good general agreement with experimental observations.

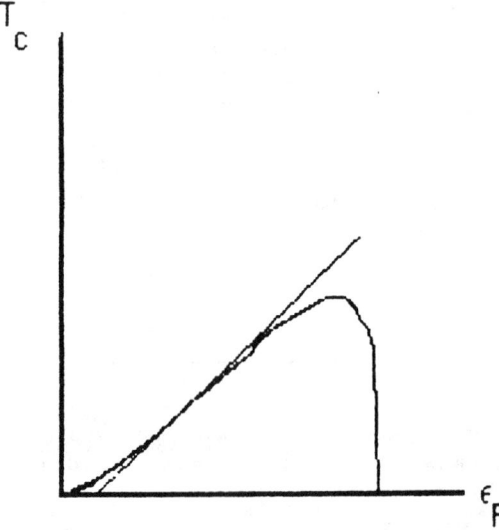

FIGURE 1 Schematic plot of Eq. (1), with Fermi energy $\epsilon_F \approx$ **n**/$m^*$ in 2D. We note that the parameter $\lambda$ is not quite independent of **n** at very small (and also at large) hole densities **m**.

With $\epsilon_F$ expressed in $eV$ and $T_c$ in $^\circ K$, (1) is: $T_c = \epsilon_F \times 13{,}120 \times e^{-1/\lambda}$, i.e. $T_c = 33\ ^\circ K$ for $\epsilon_F = 2$, $\lambda = 0.15$, and $T_c = 177\ ^\circ K$ for $\epsilon_F = 2$, $\lambda = 0.2$, spanning the range of presently known $T_c$'s.

Because two-body scattering conserves momentum, it also conserves current. Therefore, the normal-phase conductivity predicted by the present theory for $T > T_c$ is relatively high, in accord with experiment.[2] Although our mechanism does not by itself limit the normal conductivity, it may modify the electron-phonon scattering, which does.

The key questions are: how can the $e^2/r$ interaction ($4\pi e^2/q^2$ at long wavelengths in reciprocal space) *ever* be attractive? What is the role of electronic screening, which normally prevents long-range forces from being effective? What, if any, is the role of the Hubbard potential (the short-range, on-site two-body interaction $U_0$)? And why are layered materials preferred? These are all legitimate questions I shall address below. As the present theory is indifferent to magnetism (except insofar as spin-flips of $Cu^{+2}$ ions may spoil the BCS pairing mechanism), it affords hope that we shall find new high-$T_c$ superconductors based on nonmagnetic metal ions and a chemistry other than that of copper.

## TWO-BODY COULOMB FORCES—ATTRACTIVE?

It may seem paradoxical that the attractive force which binds Cooper pairs originates in the repulsive two-body Coulomb potential. Yet, the explanation is simple. Unlike point charges in the continuum, electrons (or holes) in solids are restricted to discrete lattice sites $\mathbf{R}_j$, where they occupy localized Wannier orbitals $\Phi(\mathbf{r} - \mathbf{R}_j)$. The following trivial examples on a simple-cubic lattice (with $a$ = the lattice parameter) demonstrate how spatially repulsive forces can become attractive *somewhere* in reciprocal space, even in cubic materials.

In the first, and simplest, example, along with the Hubbard repulsion parameter $U_0 = \iint d^3r d^3r' |\Phi(\mathbf{r}')|^2 |\Phi(\mathbf{r})|^2 \{e^2/|\mathbf{r} - \mathbf{r}'|\}$ we include only the nearest-neighbor repulsions $U_1 = \iint d^3r d^3r' |\Phi(\mathbf{r}')|^2 |\Phi(\mathbf{r})|^2 \{e^2/|\mathbf{r} - \mathbf{r}' + \mathbf{R}_1|\} \approx e^2/a$. Then, in reciprocal space,

$$V(\mathbf{q}) = U_0 + 2U_1(\cos q_x a + \cos q_y a + \cos q_z a). \qquad (2)$$

This exhibits a *range* of attractive $V(\mathbf{q})$ near the corners of the Brillouin Zone ("BZ") (at $\mathbf{Q} \equiv (\pm\pi, \pm\pi, \pm\pi)/a$), *unless* $U_0$ is sufficiently repulsive to exceed $6U_1$.

In our second example, we include the Coulomb force out to infinity. Because of its long range, we are, in fact, required to do so. With $U(R) = e^2/R$ (at lattice vectors $\mathbf{R} \neq 0$), and $U(0) = U_0$ as given above,

$$V(\mathbf{q}) = U_0 + \sum_{j \neq 0} (e^2/R_j) e^{i\mathbf{q} \cdot \mathbf{R}_j} \equiv U_0 + \Delta V(\mathbf{q}) \qquad (3)$$

Now, $\Delta V(\mathbf{Q}) = -1.748\, e^2/a$ (1.748 is the Madelung constant of NaCl). Thus, $V(\mathbf{q})$ *remains* attractive near $\mathbf{Q}$ *unless* $U_0$ exceeds $1.748 e^2/a$.

As our third example, we examine the potential of Eq. (3) at small $\mathbf{q} \equiv (q_x, q_y)$. (This case is the most closely related to our subsequent study of layered materials.) At small $q$ but arbitrary $q_z$, one replaces the sum in (3) by an integral, to obtain an approximate expression:[4]

$$V(\mathbf{q}, qz) \approx U0 +$$

$$(2\pi e^2/qa^2)\{[1 - C^2(q)]/[1 - 2C(q)\cos q_z a + C^2(q)] - qa/\sqrt{\pi}\} \quad (4)$$

with $C(q) \equiv \exp-(qa)$. Thus, at the point $\mathbf{q} = 0$ and $q_z = \pi/a$,

$$V(0, \pi/a) \approx U_0 + (e^2/a)\{2\pi(1/2 - 1/\sqrt{\pi})\} \approx U_0 - (e^2/a) \times 0.4033. \quad (5a)$$

Glasser[5] has independently evaluated (3) at this point, using an essentially exact summation procedure, and obtains a lower value still:

$$V(0, \pi/a) = U_0 - (e^2/a) \times 0.774386... \quad (5b)$$

Taken together, Eqs. (3) and (5) demonstrate that $V(\mathbf{q}, q_z)$ can *indeed* be negative over an extensive fraction of the outer regions of the BZ.

An attractive $V$ is a prerequisite for Cooper pairing, a necessary condition for pairing to be favored. We have now demonstrated that $V$ *can* be attractive. But is this sufficient? Hardly! In the first example above, the kernel $\mathcal{K}$ which appears in the superconducting gap equation is

$$\mathcal{K} = -\gamma^2 V(\mathbf{q}) = -\gamma^2\{U_0 + 2U_1(\cos q_x a + \cos q_y a + \cos q_z a)\}$$

with $\gamma^2$ related to the density-of-states. The seven eigenvalues $\lambda$ of $\mathcal{K}$ are: $\lambda_0 \alpha - U_0$ and $\lambda_1, \ldots, \lambda_6 \alpha - U_1$. If, as one supposes, $U_0$ and $U_1 \geq 0$, all seven are negative and, by virtue of this fact, unacceptable.

We shall find, as a result of Wannier function overlaps within each plane, that the Coulomb repulsion between *holes* located on nearest-neighbor and next-nearest-neighbor sites within any given plane drops *below* the Coulomb value $e^2/R$ at small $R$. All other things being equal, this causes the fraction of the BZ over which $V$ is attractive to grow. (Conversely, the Coulomb interaction between electron states *exceeds* $e^2/R$ and renders them unlikely candidates for high-$T_c$ superconductivity.) But it is only after self-consistent modifications due to *screening* are taken into account, that this pseudo-attraction results in a positive $\lambda$, a possible instability of holes against pairing, and ultimately, in high-temperature superconductivity.

## A MICROSCOPIC MECHANISM: COULOMB SCATTERING AMPLITUDES

Before studying the Coulomb force in rectangular materials (intralayer spacings $a$, intralayer separations $d$), and examining the decisive consequences of electronic screening, it is useful to delve into the nature and origins of the Coulomb interactions in solids. We start with the two-body scattering amplitudes, as illustrated in Fig. 2:

$$U_{k_1 k_2}(\mathbf{q})/N \equiv \iint d^3r\, d^3r'\, \Psi^*_{k_2-q}(\mathbf{r}')\Psi_{k_2}(\mathbf{r}')\{e^2/|\mathbf{r}-\mathbf{r}'|\}\Psi^*_{k_1+q}(\mathbf{r})\Psi_{k_1}(\mathbf{r}). \tag{6}$$

The $\Psi$'s are Bloch functions within the highest populated band,

$$\begin{aligned}\Psi_k(\mathbf{r}) &= e^{i\mathbf{k}\cdot\mathbf{r}} u_k(\mathbf{r}) \\ &= N^{-1/2} \sum_j e^{i\mathbf{k}\cdot\mathbf{R}_j} \Phi(\mathbf{r}-\mathbf{R}_j)\end{aligned} \tag{7}$$

and can be written either in terms of $u_k(\mathbf{r})$ (periodic from cell to cell) or of the $\Phi$'s, which are the orthonormal set of localized Wannier functions. States other than those in the highest occupied band are neglected because of their nonresonant involvement; in perturbation theory, such states would always be accompanied by a corresponding energy denominator. (In the special case that the highest band in question is *degenerate*, one must incorporate a band index in the quantum number $\mathbf{k}$ labeling the retained states. However, there is no such degeneracy in the $CuO_2$-based superconductors.[6])

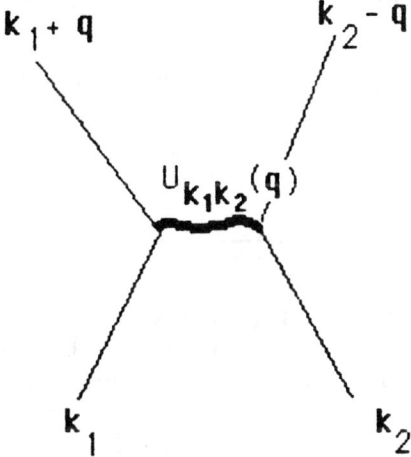

FIGURE 2 Scattering of two particles in Bloch states.

Even so, Eq. (6) is still too general for superconductivity, which requires only the *pair* scattering amplitudes (i.e. $k_2 = -k_1$),

$$V_\mathbf{k}(\mathbf{q}) \equiv U_{\mathbf{k},-\mathbf{k}}(\mathbf{q}) = \iint d^3r d^3r' \rho_\mathbf{k}^*(\mathbf{r}')\rho_\mathbf{k}(\mathbf{r}) \sum_\mathbf{R} e^{i\mathbf{q}\cdot\mathbf{R}}\{e^2/|\mathbf{r}-\mathbf{r}'+\mathbf{R}|\} \quad (8)$$

where

$$\rho_\mathbf{k}(\mathbf{r}) = \sum_j e^{i\mathbf{k}\cdot\mathbf{R}_j}\Phi^*(\mathbf{r})\Phi(\mathbf{r}-\mathbf{R}_j) \quad (9)$$

This can also be written $\rho_\mathbf{k}(\mathbf{r}) = N^{1/2}\Phi^*(\mathbf{r})\Psi_\mathbf{k}(\mathbf{r}) = \Sigma_{k'}\Psi_{\mathbf{k}'}^*(\mathbf{r})\Psi_\mathbf{k}(\mathbf{r})$. Note that $\rho_\mathbf{k}(\mathbf{r})$ has the form $|\Phi(\mathbf{r})|^2 + \Delta_\mathbf{k}(\mathbf{r})$, where $\Delta_\mathbf{k}(\mathbf{r})$ is an oscillating probability distribution centered on the origin, the average of which is zero (by orthogonality of the Wannier states.) We rewrite (8) in a way which highlights the spatial dependence of the interaction energy:

$$V_\mathbf{k}(\mathbf{q}) = \sum_\mathbf{R} e^{i\mathbf{q}\cdot\mathbf{R}} w_\mathbf{k}(\mathbf{R}) \quad (10)$$

with

$$w_\mathbf{k}(\mathbf{R}) \equiv \iint d^3r d^3r' \rho_\mathbf{k}^*(\mathbf{r}')\rho_\mathbf{k}(\mathbf{r})\{e^2/|\mathbf{r}-\mathbf{r}'+\mathbf{R}|\} \quad (11)$$

Asymptotically, $w_\mathbf{k}(\mathbf{R}) \to e^2/R$ at all $\mathbf{k}$ (proof: neglect $\mathbf{r}-\mathbf{r}'$ at large $\mathbf{R}$, use (9) and orthonormality of $\Phi$'s.) At $\mathbf{R} = 0$, we define a function of $\mathbf{k}$,

$$W_0(\mathbf{k}) \equiv w_\mathbf{k}(0) = \iint d^3r d^3r' \rho_\mathbf{k}^*(\mathbf{r}')\rho_\mathbf{k}(\mathbf{r})\{e^2/|\mathbf{r}-\mathbf{r}'|\}. \quad (12)$$

This function of $\mathbf{k}$ is the appropriate generalization of the Hubbard parameter $U_0$.

## WHY ARE HOLES PREFERRED?

We show, qualitatively and to a first approximation, that the interaction $w_\mathbf{k}(\mathbf{R})$ is more repulsive for low-lying states than for the high-lying states in the band (those occupied by holes); the particle-hole symmetry is thus manifestly broken by the Coulomb force.

For the sake of argument, assume that the nonvanishing overlaps in (9) occur principally between nearest-neighbor sites (and, for the layered materials, principally for those which are also coplanar).

Then, consider the nearest-neighbor overlap integral:

$$t_j \equiv \int d^3r \Phi^*(\mathbf{r})\{-e^2[1/r + 1/|\mathbf{r} - \mathbf{R}_j|]\}\Phi(\mathbf{r} - \mathbf{R}_j) \quad (13)$$

assuming the ions all have charge $+e$; otherwise replace $-e^2$ by the corresponding $-ze^2$. In tight-binding theory, the Fourier transform of $t_j$ yields the Bloch energies $e(\mathbf{k})$. For layered materials with lattice parameters $a$ in each plane and $d$ perpendicular, these are:

$$e(k_x, k_y, k_z) = 2(e^2/a)\tau_x(\cos k_x a + \cos k_y a) + 2t_z \cos k_z d, \quad (14)$$

defining $\tau = ta/e^2$. Assuming the overlap between layers to be negligible, $|t_z| << |t_x|$, the surfaces of constant energy are quasi-cylindrical (as shown in Fig. 3). We can therefore *crudely* estimate (11):

$$w_\mathbf{k}(R) \approx \{e^2/R\}\{1 - 2\tau_x(\cos k_x a + \cos k_y a)\}$$
$$\approx \{e^2/R\}\{1 - \theta(k_x, k_y, k_z)(a/e^2)\} \quad (15)$$

for $R \neq 0$. The highest/lowest values of the Bloch energy, $\theta_\pm \approx \pm 4(e^2/a)|\tau_x|$, correspond to hole/electron states. Thus, according to (15), $w_\mathbf{k}(R)$ has its maximum/minimum value: $\{e^2/R\}\{1 \pm 4|\tau_x|\}$, for electrons/holes.   QED

The analysis of Eq. (15) cannot be used at $R = 0$, nor is there any $R$ at which it is rigorous. Nevertheless, it provides a preliminary indication that the two-body potential between particles at the bottom of the band (*electrons*), spatially separated by $R$, exceeds the estimate $\iint d^3r d^3r' |\Phi(\mathbf{r}')|^2 |\Phi(\mathbf{r})|^2 \{e^2/|\mathbf{r} - \mathbf{r}' + \mathbf{R}|\} \approx e^2/R$, while, conversely, the two-body repulsion between particles in states near the top of the band (*holes*) falls short of it. Still, all corrections vanish at $R \to \infty$ (as proved earlier), and $w_\mathbf{k}(R) \to e^2/R$ asymptotically at all values of $\mathbf{k}$.

## BARE COULOMB INTERACTIONS

To take into account the destructive interference between $|\Phi(\mathbf{r})|^2$ and $\Delta_\mathbf{k}(\mathbf{r})$ for states near the top of the band, *we simply replace $e^2/R$ by $U(\mathbf{R}) \equiv e^2/[R^2 + a_w^2]^{1/2}$, for all $\mathbf{R}$ within* the plane, with $a_w$ defined to be the radius of the Wannier orbital. This correction reflects in a semiquantitative way, the erosion of the Coulomb repulsion between holes at finite distances $R$. These corrections vanish asymptotically:

$$[U(R) - e^2/R] \sim -e^2 a_w^2/2r^3,$$

as they should. Insofar as $t_z$ is small, *it is unnecessary* to make any similar adjustments for two particles on *different* planes.

I have found it convenient to relate $a_w$ to the Bohr radius appropriate to the given band structure, $a_0^* \equiv \hbar^2/m^* e^2$, by $a_w = \alpha a_0^*$. The resulting parameter $\alpha$ is the ratio of Wannier orbital radius to Bohr orbit radius, the former being a distance characteristic of short-range forces (e.g., the Hubbard interaction parameter $U_0 \approx e^2/a_w$), and the latter being characteristic of the particular band structure. While $a_w$ is typically $O(a)$, the Bohr radius can vary from a fraction of an Ångstrom to a large multiple of the lattice parameter, depending on the precise value of $m^*$. Thus, $\alpha$ could vary greatly with the material properties. We shall see that the pairing mechanism materially depends on whether $\alpha$ lies in the vicinity $\alpha \approx 1$, or in the range $\alpha > 2$, especially near $\alpha = 5$.

Considering forces within a single plane only, we can obtain the two-dimensional Fourier transform in closed form by approximating the lattice sum with a two-dimensional integral,

$$V(q) \equiv (e^2/a) \sum_{(n,m)} e^{ia(q_x n + q_y m)} [n^2 + m^2 + (a_w/a)^2]^{-1/2} \tag{16}$$

$$\approx (e^2/a^2) \int dx \int dy \, e^{i(q_x x + q_y y)} [x^2 + y^2 + a_w^2]^{-1/2} = \{2\pi e^2/qa^2\} e^{-qa_w}$$

The sum on the first line and the integral which is evaluated on the second line have been independently compared by H.-Q. Lin,[7] who finds their difference to be insignificant at long wavelengths ($qa \leq 1$) for $a_w \geq a$.[8]

Incorporating the straight Coulomb *inter*planar interactions, the *three*-dimensional Fourier transform is:

$$U(\mathbf{q}, q_z) = \sum_{n,m} \sum_{s \neq 0} e^{i(aq_x n + aq_y m + dq_z s)} e^2/[a^2 n^2 + a^2 m^2 + d^2 s^2]^{1/2} \tag{17}$$

$$+ \sum_{n,m} e^{i(aq_x n + aq_y m)} e^2/[a^2 n^2 + a^2 m^2 + a_w^2]^{1/2}$$

At each $s$ the two-dimensional sum on $(n, m)$ is approximated by the corresponding integral, without substantial error.[7] The sum over $s$ has the form of a geometric series, which can be evaluated without approximation. At small $q = \sqrt{(q_x^2 + q_y^2)}$ but at arbitrary $q_z$, the result is:

$$U(\mathbf{q}, q_z) = (A/q)\{[1 - B^2(q)][1 - 2B(q)\zeta + B^2(q)]^{-1} - \Gamma(qa_w)\} \tag{18}$$

in which $A \equiv 2\pi e^2/a^2$, $B(q) \equiv \exp(-(qd))$, $\zeta \equiv \cos(q_z d)$, and $\Gamma$ contains the corrections to the Coulomb forces,

$$\Gamma(qa_w) = 1 - e^{-qa_w} = qa_w - (qa_w)^2/2 + \cdots \tag{19}$$

If $q$, $q_z$ are both small, (18) reduces to the "textbook" result for "jellium": $U(\mathbf{q}, q_z) \to (4\pi e^2/a^2 d)(q^2 + q_z^2)^{-1}$, where $a^2 d$ = volume of unit cell. In this limit, $U(\mathbf{q}, q_z)$ is indeed positive definite and devoid of any intrinsic possibilities for superconductivity. Indeed, for $d >> a_w$, $U(\mathbf{q}, q_z)$ *remains* repulsive at all $q_z$. But while it might appear that the correction $\Gamma$ is somewhat academic in that case, *even there* the textbook jellium formula is invalid, except at long wavelengths $q, q_z \to 0$.

As a function of $q_z$, $U(\mathbf{q}, q_z)$ has its minimum at $q_z = \pi/d$ ($\zeta = -1$). There it is negative whenever $d/2a_w \leq 1$. Fig. 3b illustrates the consequences of this for scatterings on quasi-cylindrical Fermi surfaces.

## SCREENING AND THE DIELECTRIC FUNCTION

The dielectric function $\epsilon(\mathbf{q}, q_z; \omega)$ supplies the necessary corrections to the bare potentials, accounting for the current flow and polarization of the electronic fluid about each carrier.

In thermal equilibrium, a charged fluid cannot sustain an electric field over any macroscopic distance. Thus charged particles become neutralized beyond a "screening distance" and the interaction between two such particles vanishes beyond this distance. While the screening mechanism is difficult to describe in real space, it is relatively simple in reciprocal space within the Lindhard or RPA formalisms.[9] The two-body interactions are required to be spectrally decomposed, each Fourier component then being divided by the dielectric function $\epsilon(\mathbf{q}, q_z; \omega)$. Here, $\mathbf{q}, q_z, \omega$ are the momenta and energy transferred at a vertex (Fig. 2), with $\omega$ a function of momentum transfer $\mathbf{q}, q_z$.

Consider only a low-density gas of carriers at first, for simplicity. At small $q$ the dielectric function of a layered material with a two-dimensional band structure is[6] significant[10] *even at low densities*:

$$\epsilon(\mathbf{q}, q_z; \omega) = 1 + \mathbf{X}(z, \nu)\alpha u(\zeta) \tag{20}$$

where

$$u(\zeta) = [2F(1 - \zeta)^{-1} - 1] \tag{21}$$

is extracted from (18) in the limit $q \to 0$. Eq. (21) incorporates a key parameter $F \equiv d/2a_w$, and $\mathbf{X}$ is the complex quantity:

$$\begin{aligned}\mathbf{X} &= R - iI \\ &= 1 - (2z)^{-1}C_{-}[|(z - \nu)^2 - 1|^{1/2} - (2z)^{-1}C_{+}[|(z + \nu)^2 - 1|^{1/2}\end{aligned} \tag{22}$$

with $C_{\pm} = 1$ if $(z \pm \nu)^2 > 1$ and $C_{\pm} = i$ otherwise.

Then, the appropriately retarded, causal, screened, pair potential $V$ which enters the BCS formulation in the $q \approx 0$ limit is $V = \text{Re}\{U/\epsilon\}$:

$$V(z, \zeta, \nu) = Aa_0^*\alpha u(\zeta)[1 + R(z, \nu)\alpha u(\zeta)]/$$
$$\{[1 + R(z, \nu)\alpha u(\zeta)]^2 + [I(z, \nu)\alpha u(\zeta)]^2\} \quad (23)$$

in which the variables are: $z = |\mathbf{k} - \mathbf{k'}|/2k_F$, $\zeta = \cos(k_z - k_z')d$, and $\nu = |\mathbf{k} + \mathbf{k'}|/2k_F$, appropriate for the scattering of one of the paired quasi-particles from $\mathbf{k}$ to $\mathbf{k'}$.

These results are exact (within RPA) but cumbersome. We simplify them somewhat by noting that the typical particle is at or near $k_F$. The parameter governing $z$, $\nu$ is just the angle $\varphi = \cos^{-1}\{\mathbf{k}\cdot\mathbf{k'}/|k||k'|\}$. Setting $|k| \approx |k'| \approx k_F$ in (19), we obtain $X = 1 - (1 + i)[(1/2)|\cot \varphi/2|]^{1/2}$. This result is, remarkably, independent of $k_F$, hence of carrier density (except possibly at the very lowest densities, where the band structure at small $\mathbf{k}$ becomes three-dimensional—as in the example of Fig. 3a).

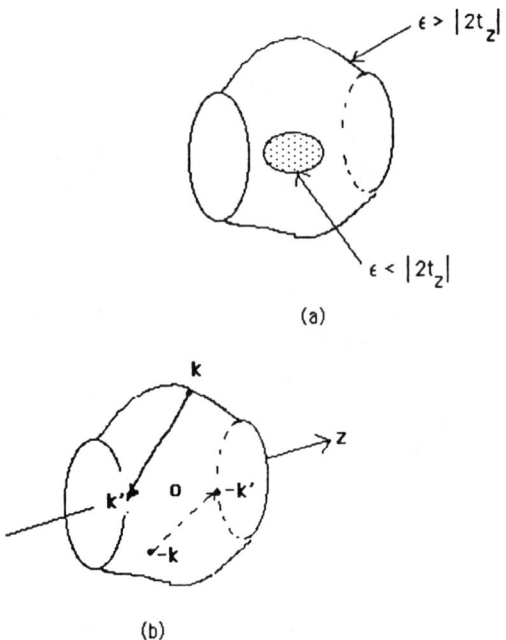

FIGURE 3 (a) 2D constant-energy surfaces are quasi-cylindrical, but become ellipsoidal and quasi-three-dimensional very close to band edges (when $2|t_z|$ becomes significant compared with total Bloch energy $\epsilon$).

(b) Very large-momentum scatterings $q_z = k_z - k_z'$—as large as $q_z = \pi/d$—are allowed on the quasi-cylindrical Fermi surface, as shown, but not on an ellipsoidal surface. According to Eqs. (18) and (27), the corresponding scattering amplitudes can be negative (attractive), hence conducive to Cooper-pairing and superconductivity.

At this point, anticipating that the pairing function will not depend on $\varphi$, we replace $1/\epsilon$ by $\langle \mathrm{Re}\{1/\epsilon\}\rangle$, the average being over the angle $\varphi$. The correct screened interaction to be considered in the pairing problem of a low-density gas is then:

$$V_{scr}(\zeta) = Aa_0^* \alpha u(\zeta) \times (1/\pi) \int_0^\pi d\varphi [1 + \alpha u(\zeta) R(\varphi)] /$$

$$\{[1 + \alpha u(\zeta) R(\varphi)]^2 + [\alpha u(\zeta) I(\varphi)]^2\} \quad (24)$$

where $R(\varphi) = 1 - [(\cot \varphi/2)/2]^{1/2}$ and $I(\varphi) = -[(\cot \varphi/2)/2]^{1/2}$.

In Fig. 4 (inset) we show $\alpha u(\zeta)$, the bare interaction of Eq. (18) at small $q$, in units of $Aa_0^*$. With $\zeta = \cos q_z d$, we plot this potential over the interval $0 \leq q_z d/\pi \leq 1$ (dashed curve), taking for our example $F = 0.25$ and $\alpha = 1$, comparing it to the corresponding screened potential of Eq. (24) (in the same units). The unscreened is far more repulsive than the screened interaction at small $q_z$ (indeed, the former *diverges* as $q_z^{-2}$ in the limit $q_z \to 0$). In Fig. (5) (inset) the corresponding bare and screened potentials at $\alpha = 5$ and $F = 0.8$, are illustrated. Note that the $q_z$ divergence is eliminated in all cases.

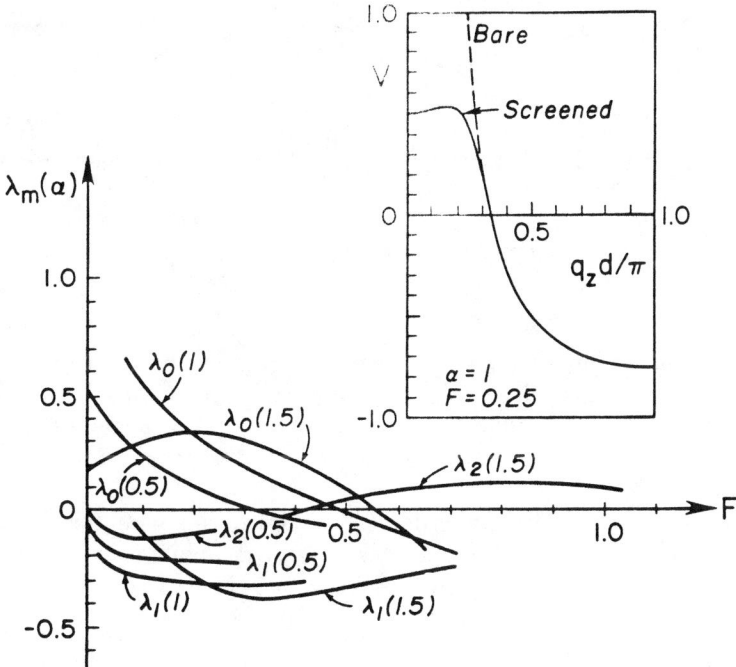

FIGURE 4 Eigenvalues $\lambda_m$ vs. $F = d/2a_w$ for $\alpha = 0.5, 1, 1.5$ and various $m$. Only $\lambda_0$ is sufficiently large (and $>0$) to yield superconductivity in this region of $\alpha$. *Inset:* Bare and screened potentials $V(0, q_z)$ in units of $1/Aa_0^*$.

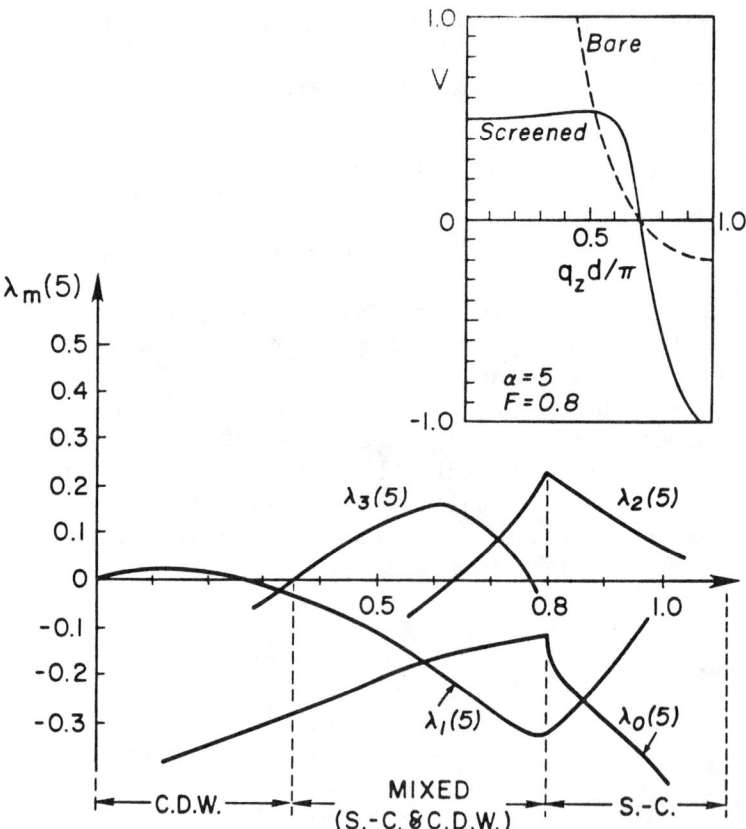

FIGURE 5 Eigenvalues $\lambda_m$ vs. $F$ at $\alpha = 5$ (the optimal $\alpha$ for $\lambda_2$). Regions of instability against charge-density-waves are also shown. *Inset:* bare and screened potentials.

The present formulation ignores plasma resonances (found at the zeros of $\epsilon$). While it is possible that the virtual exchange of plasmons, excitons, magnons, phonons, etc., all provide additional sources of attraction, none of these mechanisms are essential to the present development, and they are left to be examined separately.

We next introduce the dimensionless screened pair-potential function $\mathcal{H} \equiv -V \times$ two-dimensional density of states, $N(e)$. (Assuming $e_F > |2t_z|$, the Fermi surface is quasi-cylindrical as shown in Fig. 3 and the effective two-dimensional density-of-states function is approximately a constant $N(e) = N(e_F) = 1/Aa_0^*$.) Thus,

$$\mathcal{H}(\zeta) = -\alpha u(\zeta)(1/\pi) \int_0^\pi d\varphi [1 + \alpha u(\zeta) R(\varphi)]/$$

$$\{[1 + \alpha u(\zeta) R(\varphi)]^2 + [\alpha u(\zeta) I(\varphi)]^2\} \quad (25)$$

(note the "−" sign) is the desired quantity. Recall, $R(\varphi) = 1 - [(\cot \varphi/2)/2]^{1/2}$ and $I(\varphi) = -[(\cot \varphi/2)/2]^{1/2}$.

$\mathcal{K}(\zeta)$ will turn out to be the kernel of the BCS integral equation. For superconductivity to occur, at least one of its eigenvalues must be positive. Where the bare potential $\alpha u(\zeta)$ diverged at $\theta$ or $q_z \to 0$, $\mathcal{K}$ now approaches a finite, albeit negative, limit. At the other limit, $q_z d = \pi$, $\mathcal{K}$ can become positive if $F < 1$.

The relevance of electronic screening can be assessed as follows: with $I = R \equiv 0$, (25) reduces to $\mathcal{K}_{\text{bare}} = -\alpha[(2F)/(1 - \zeta) - 1]$. Owing to the singularity at $\theta = 0$ all eigenvalues of this bare kernel are either $-\infty$ or 0, and *none* are positive. Thus, the long-range of the unscreened Coulomb interaction renders it inimical to pairing, and to superconductivity, regardless *whether or not* there are attractive regions in **q** space.

## FINITE CARRIER DENSITY

To observe the drop-off in $T_c$ at high density, as shown in Fig. 1, we calculate the kernel at finite $k_F$. It is convenient to define $a_w k_F \equiv \kappa_F$, with $a_w$ turning out to be the "effective" lattice parameter in the plane (rather than $a$). The density **n** of charge carriers is,

$$\mathbf{n} = 2\epsilon_F/Aa_0^* = (1/2\pi)(a/a_w)^2 \kappa_F^2, \tag{26}$$

$$V_{\text{bare}} = (A\alpha a_0^*/qa_w)\{[1 - B^2]/[1 - 2\zeta B + B^2] - \Gamma(qa_w)\} \tag{27}$$

following from (18). With $\varphi$ defined following (23), we eliminate $q$:

$$q = 2k_F|\sin \varphi/2| \tag{28}$$

and obtain the dielectric function as:

$$\epsilon(\zeta, \varphi) = 1 + \mathbf{X}(\varphi)\alpha u(\zeta, \varphi) \tag{29}$$

Here $u(\zeta, \varphi)$ generalizes the former $u(\zeta)$ of Eq. (21), and is:

$$u(\zeta, \varphi) = (2\kappa_F|\sin \varphi/2|)^{-1}\{[1 - B^2]/[1 - 2\zeta B + B^2] - \Gamma(qa_w)\} \tag{30}$$

Obviously, $u(\zeta, 0) = u(\zeta)$. With $F \equiv d/2a_w$, the quantities $B$, $\Gamma$ are now

$$B(q) \equiv \exp-(4F\kappa_F|\sin \varphi/2|)$$
$$\Gamma(qa_w) \equiv 1 - \exp[-(2\kappa_F|\sin \varphi/2|)] \tag{31}$$

and $\zeta \equiv \cos \theta = \cos k_z d$, as before. Recall:

$$\mathbf{X} \equiv R - iI = 1 - (1 + i)|(1/2)\cot \varphi/2|^{1/2} \tag{32}$$

Then replacing (25), the kernel at finite density is:

$$\mathcal{K}(\zeta) = -\int_0^\pi (d\varphi/\pi)\alpha u(\zeta, \varphi)[1 + R(\varphi)\alpha u(\zeta, \varphi)]/D^2 \quad (33)$$

where

$$D^2 \equiv [1 + R(\varphi)\alpha u(\zeta, \varphi)]^2 + [I(\varphi)\alpha u(\zeta, \varphi)]^2. \quad (34)$$

The new independent parameter is $\kappa_F$, supplementing $\alpha$ and $F$. When it is small (e.g. $\leq 0.1$) we recover previous results. When it is not, the magnitude of $|u|$ decreases, i.e. *effectively* $\alpha$ decreases.

## PAIRING HAMILTONIAN

Defining pair operators $b^*(\mathbf{k}, k_z) \equiv c^*(\mathbf{k}, k_z, \uparrow)c^*(-\mathbf{k}, -k_z, \downarrow)$ in terms of the Bloch operators,[1] the effective Hamiltonian satisfied by the pairs in the absence of current is:

$$H = \sum (e(\mathbf{k}, k_z) - e_F)c^*(\mathbf{k}, k_z, \uparrow)c(\mathbf{k}, k_z, \uparrow)$$
$$+ (1/NN_z) \sum V(z, \zeta, \nu)b^*(\mathbf{k}, k_z)b(\mathbf{k}', k_z') \quad (35)$$

The gap equation which ensues is one of the well-known equations in solid state physics:[1]

$$\Delta(\mathbf{k}, k_z, T) = -(1/2NN_z) \sum_{\mathbf{k}'} V(z, \zeta, \nu)\Delta(\mathbf{k}', k_z', T)E(\mathbf{k}', k_z')^{-1} \tanh \beta E(\mathbf{k}', k_z')/2 \quad (36)$$

with $\beta = 1/kT$, $E(\mathbf{k}, k_z) \equiv [(e(\mathbf{k}, k_z) - e_F)^2 + |\Delta(\mathbf{k}, k_z, T)|^2]^{1/2}$, and $V(z, \zeta, \nu)$ given in Eq. (27).

## THE ENERGY-GAP EQUATION

Changing sums to integrals in (36), we next derive the BCS integral equation,[1] recalling that $\zeta$ in the kernel $\mathcal{K}(\zeta)$ relates momentum transfer in the $z$-direction, $\zeta = \cos(\theta - \theta')$ with $\theta = k_z d$ and $\theta' = k_z' d$. It is:

$$\Delta(\theta, T) = \int d\theta'/2\pi \mathcal{K}(\cos(\theta - \theta'))\Delta(\theta', T)\mathcal{L}(\theta'; T, t_z, e_F) \quad (37)$$

where

$$\mathcal{L}(\theta; T, t_z, e_F) \equiv \int_0^{e_F - 2t_z \cos\theta} dx (E)^{-1} \tanh \beta E/2 \quad (38)$$

with $x = \hbar^2(k^2 - k_F^2)/2m^*$, and $E \equiv [x^2 + |\Delta(\theta, T)|^2]^{1/2}$.

To the extent that the solution $\Delta(\theta, T)$ is a nontrivial function of $\theta$, various experiments will reveal different averages. One of the goals in studying the integral equation is to obtain the distribution $P(\Delta)$ of $\Delta$ in phase space.

If the parameter $t_z$ is set equal to zero, the integral equation becomes trivially soluble upon setting $\Delta(\theta, T) = \Delta_0(T)\exp(im\theta)$, with $m$ determined by the largest (positive) eigenvalue $\lambda_m$ of $\mathcal{K}$ (see below).

But if $t_z$ vanished indeed, the paired states become subject to the giant fluctuations (and lack of long-range order) characteristic of any condensed phase of continuous symmetry in 2D, as mandated by the Mermin-Wagner theorem.[11] Two-dimensional-type fluctuations *have* been observed, but only over a narrow range of temperatures near $T_c$.[12] But for 3D superconductivity, one needs a finite $t_z$, with all the attendant complications in the integral equation (mixing of eigenvalues, etc). As $t_z$ is increased and the conductivity and band structure become more three-dimensional, we observe $T_c$ to decrease substantially.

Noting that $\mathcal{K}(\cos(\theta - \theta'))$ is real periodic, we expand it as follows:

$$\mathcal{K}(\cos(\theta - \theta')) = \sum_m \lambda_m e^{im(\theta - \theta')} \quad (39)$$

with $\lambda_m = \lambda_{-m} = \lambda_m^*$. To date, our numerical calculations have been limited to the low-density limit. We plot $\lambda_0, \lambda_1, \lambda_2, \lambda_3$ as functions of $F$ and $\alpha$ in this limit, in the accompanying Figs. 4 ($\alpha \approx 1$), Fig. 5 ($\alpha = 5$, where $\lambda_2$ achieves its optimum value), and Fig. 6, where the optimum (maximum) $\lambda$ at each $\alpha$ is plotted over the range $0 \leq \alpha \leq 9$, together with the corresponding value of $F$ at which this optimum is achieved.

Now we rewrite Eq. (37) by defining $\psi(\theta, T) \equiv \Delta(\theta, T)\mathcal{L}(\theta)$, and using the expansion (39), obtain:

$$\psi(\theta) = \mathcal{L}(\theta) \sum_n \lambda_n \xi_n e^{in\theta}, \quad \text{where} \quad \Delta(\theta) = \sum_n \lambda_n \xi_n e^{in\theta}, \quad (40)$$

where the Fourier coefficients $\xi_n$ are:

$$\xi_n \equiv \langle e^{-in\theta} \psi(\theta) \rangle_\theta, \quad \text{or inversely,} \quad \psi(\theta) = \sum_n \xi_n e^{in\theta}, \quad (41)$$

defining $\theta$-averages as: $\langle F(\theta) \rangle_\theta \equiv \int_{-\pi}^{+\pi} F(\theta) d\theta/2\pi$.

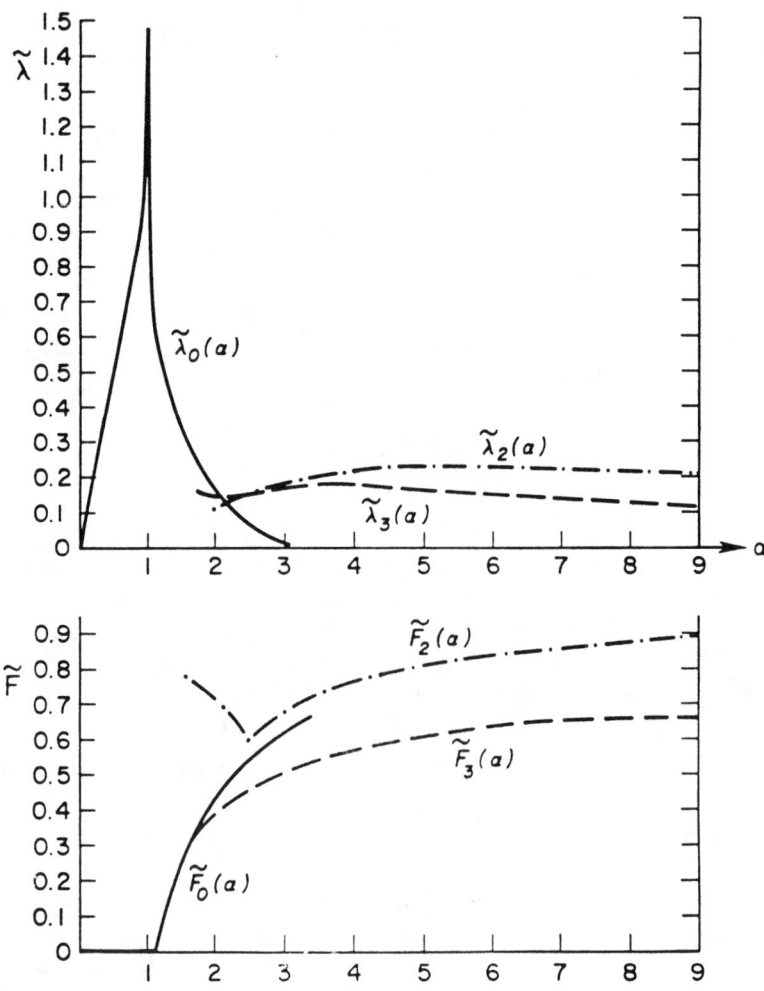

FIGURE 6 Upper curves; maximum eigenvalues $\tilde{\lambda}_m(\alpha)$ vs. $\alpha$. Lower curves: $\tilde{F}(\alpha)$, the value of $F = d/2a_w$ at which the optimal $\lambda$ is found.

The $k_2$-dependence of the gap function is explicit in (40), with $\Delta(\theta) = \lambda_0\xi_0 + 2\lambda_1\xi_1 \cos \theta + 2\lambda_2\xi_2 \cos 2\theta + \cdots$ being a real quantity only if $\lambda_0$ is the leading (positive) eigenvalue. Once $\alpha \geq 3$ and $\lambda_2$ becomes the leading eigenvalue, the expansion (40) will center either about $n = +2$ (i.e. $\lambda_2\xi_2 e^{2i\theta} + \lambda_1\xi_1 e^{i\theta} + \lambda_3\xi_3 e^{3i\theta} + \cdots$) or about $n = -2$, and $\Delta$ becomes *inherently* complex.

The $\lambda_n$ are the eigenvalues of the kernel, are calculated directly, and shown in the figures. The $\xi_n$ are functions of $T$, and must be found as the self-consistent solutions of the nonlinear coupled Eqs. (40) and (41). The expansion may be truncated when either or both these coefficients are deemed negligible.

## CALCULATION OF $T_c$ WHEN $\lambda_0$ IS PREDOMINANT ($\alpha \leq 2$)

At $T_c$ only, Eqs. (40) and (41) can be solved simultaneously by ordinary linear algebra. There, $\mathscr{L}$ is given approximately as follows:

$$\mathscr{L}(\theta; T_c, t_z, e_F) \equiv \int_0^{e_F - 2t_z \cos\theta} dx(x)^{-1} \tanh \beta_c x/2$$
$$\approx \ln\{1.13(1/kT_c)(e_F - 2t_z \cos\theta)\}, \quad (42)$$

accurate at all $\theta$ provided $kT_c$ is not too large and $|2t_z| < e_F$. It is now convenient to introduce a dimensionless quantity $\tau \equiv 2|t_z|/e_F$ with which to further simplify (42). For the moment assume the parameter $\tau$ to be in the easily analyzed range $\tau < 1$.

If $\lambda_0$ is most important in the expansion of the kernel, as is generally the case for $\alpha \leq 2$, Eqs. (40) and (41) yield

$$\psi(\theta) = \mathscr{L}(\theta)\lambda_0\xi_0, \quad \text{i.e.} \quad \langle\psi(\theta)\rangle_\theta = \xi_0 = \lambda_0\xi_0\langle\mathscr{L}(\theta)\rangle_\theta \quad (43)$$

This last either has the trivial solution $\xi_0 = 0$, or else, $\lambda_0\langle\mathscr{L}(\theta)\rangle_\theta = 1$. Performing the angular average on (42) we find:

$$kT_c = (1.13 e_F)\{1/2 + [1 - \tau^2]^{1/2}/2\}\exp(-1/\lambda_0) \quad (44)$$

for the nontrivial solution. This shows that at constant $e_F$, $T_c$ decreases by as much as 50% when $\tau$ is increased from $\tau = 0$ to 1.

When $\tau > 1$ the approximation (42) for $\mathscr{L}(\theta; T_c, t_z, e_F)$ becomes invalid, because the topology of the Fermi surface changes from a distorted cylinder to an ellipsoid, as shown in Fig. 3. We shall now follow the calculation into the three-dimensional region, beyond noting that once the energy surfaces become ellipsoidal the dielectric function weakens, as does the kernel (see remarks following Eqs. (33) and (34)), and the formalism changes completely.

It is possible to enlarge the parameter space to include two or more eigenvalues,[4] but the algebra becomes correspondingly more elaborate. Assuming only $\lambda_0$ and $\lambda_1$ need be retained, (40) implies:

$$\Delta(\theta) = \lambda_0\xi_0 + 2\lambda_1\xi_1 \cos\theta \quad (45)$$

while (41) yields:

$$\xi_n = \langle\cos n\theta \, \mathscr{L}(\theta)\Delta(\theta)\rangle_\theta \quad (46)$$

At $T = 0$, provided only that $|\Delta| < e_F/3$ and $\tau < 1$, $\mathscr{L} \equiv \int_0^{e_F - 2t_z \cos\theta} dx(x^2 + \Delta^2(\theta))^{-1/2}$ is accurately approximated by:

$$\mathscr{L}(\theta; 0, t_z, e_F) \approx \ln\{(2e_F)(1 - \tau \cos\theta)/\Delta(\theta)\} \quad (47)$$

The coupled, nonlinear, transcendental equations (45), (46) now become exactly soluble; we refer the reader elsewhere for the details.[4]

Our conclusions, to leading order in $\tau$, include:

$$\Delta_0 = 2e_F \exp(-1/\lambda_{0\text{eff}}) \qquad (48)$$

which defines the "effective" coupling constant $1/\lambda_{0\text{eff}}$. A similar correction must be made to $T_c$. Although the renormalization of $\lambda$ can be substantial, the principal effect is on $\Delta(\theta)$,

$$\Delta(\theta) \approx \Delta_0(1 - e_z \cos\theta) \qquad (49)$$

in leading order of the expansion in powers of $\tau$. At small $|\tau|$, $e_z = (4t_z/e_F)(1 + 2/\lambda)^{-1}$, possibly small, but not identically zero.

## GAP- AND QUASIPARTICLE-DISTRIBUTIONS

According to (49), $\Delta(\theta)$ lies between a maximum and a minimum $\Delta_0(1 \pm e_z)$. The probability density of finding $\Delta$ at a value within this range is:

$$P(\Delta) = (1/\pi)[(\Delta_0(T)(1 + e_z) - \Delta)(\Delta - \Delta_0(T)(1 - e_z))]^{-1/2} \qquad (50)$$

with the probability density $P$ normalized to $\int d\Delta P(\Delta) = 1$ in the usual way.

Experiments which measure $\Delta(\theta)$ differently yield, perforce, distinct numbers. The average gap is $\Delta_0 \equiv \int d\Delta \Delta P(\Delta)$, by (49).

In NMR experiments, $\Delta_{\min}$ is singled out by $1/T_1 \propto \exp[-(|\Delta_{\min}|/kT)]$ at $T << T_c$. The results of M. Lee et al[13] on LSrCO are: $2\Delta_{\min}/kT_c \approx 1.3$. With $2\Delta_0/kT_c = 3.54$ and $\Delta_{\min} = \Delta_0(1 - e_z)$, NMR thus yields $e_z \approx 0.63$.

Tunneling and optical spectra are sensitive to the distribution $\rho(E)$ of quasiparticle energies $E = \pm[(e(\mathbf{k}, k_z) - e_F)^2 + \Delta^2(k_z)]^{1/2}$. This function is easily determined to be:

$$\rho(E) = N(e_F) \, E \, \langle [E^2 - \Delta^2(\theta)]^{-1/2} \rangle_\theta, \qquad (51)$$

a form of elliptic function, vanishing in the range $-\Delta_0(1 - e_z) < E < +\Delta_0(1 - e_z)$, increasing thereafter to a logarithmic singularity at $E = \pm\Delta_0(1 + e_z)$, then decreasing to $N(e_F)$ at $|E| >> \Delta_0$. This general behavior is illustrated in Fig. 7, where the distribution of gaps $P(\Delta)$ is also shown. Optical absorption or reflection should be particularly sensitive to the peak in the density of states.

With $\lambda_2$ as the dominant positive eigenvalue for $\alpha > 2$, $\Delta$ becomes complex,

$$\begin{aligned}\Delta(\theta) &= \lambda_2\xi_2 e^{2i\theta} + \lambda_1\xi_1 e^{i\theta} + \lambda_3\xi_3 e^{3i\theta} + \cdots \\ &= \Delta_0 e^{2i\theta}(1 + e_1 e^{-i\theta} + e_3 e^{i\theta} + \cdots)\end{aligned} \qquad (52)$$

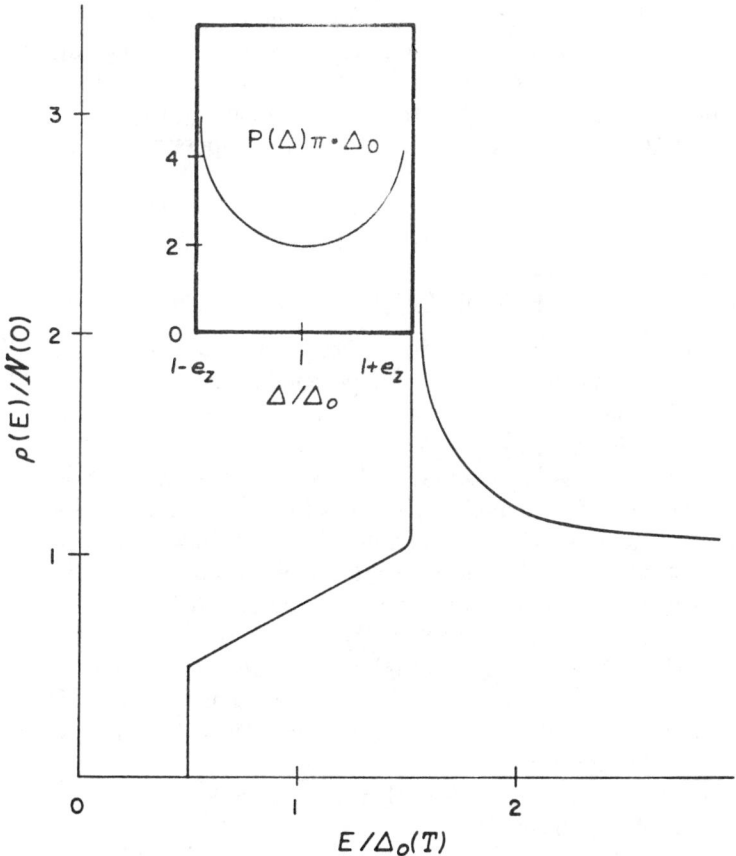

FIGURE 7 Typical quasi-particle density-of-states function $\rho(E)$ at $T = 0$. Inset: gap distribution $P(\Delta)$, assuming a typical $e_z = 0.5$.

The gap distribution function $P(|\Delta|)$ remains approximately, if not precisely, of the form graphed in Fig. 7—as does the quasi-particle density-of-states $\rho(E)$. $T_c$ continues to be given by Eq. (1), now with $\lambda_2$ (or, after $\tau$ is properly taken into account, with a corrected value $\lambda_{2\text{eff}}$).

The ground-state degeneracy $m = +2$ or $-2$ is an interesting feature in this instance, possibly with observable consequences for those superconductors having parameter $\alpha$ in the neighborhood of $\alpha \approx 5$.

## ACKNOWLEDGMENTS

Dr. H.-Q. Lin (Brookhaven National Laboratory) has been most helpful in checking the accuracy of lattice Fourier transforms. Prof. M. L. Glasser (Clarkson

University) kindly supplied the result quoted in Eq. (5b). Dr. A. M. Szpilka (University of Utah) evaluated the eigenvalues $\lambda$ plotted in the various figures, correcting and extending my early calculations. I am grateful to them for this indispensable help, and to the Electronics Technology and Devices Laboratory, SLCET-E, ARO (Ft. Monmouth, N.J.) for partial support.

## FOOTNOTES & REFERENCES

1. J. Bardeen, L. Cooper and R. Schrieffer, Phys. Rev. **108**, 1175 (1957). For a review of theory and practice of superconductivity theory (both BCS and Ginzburg-Landau) see the text of M. Tinkham, "Introduction to Superconductivity," McGraw-Hill, New York, 1975.
2. The traditional (low-$T_c$) superconductors such as mercury, tin, lead, . . . are generally poor metals above $T_c$. The opposite is true for the high-$T_c$ materials, as pointed out in M. Gurvitch and A. T. Fiory, Phys. Rev. Lett. **59**, 1337 (1987). They find the electrical resistivity in the normal phases of YBaCuO and LaSrCuO, attributed to phonon scattering, to be low—too low, in fact, to be compatible with the phonon-exchange mechanism of superconductivity. By extension, their work may also be seen to preclude the exchange of any other intermediate boson: exciton, plasmon, magnon, etc., any of which would similarly have curtailed the normal conductivity below the experimentally observed values.
3. Y. J. Uemura et al., Phys. Rev. **B38**, 909 (1988).
4. D. C. Mattis, unpublished.
5. M. L. Glasser (private communication).
6. L. Mattheiss, Phys. Rev. Lett. **58**, 1028 (1957).; J. Yu, A. Freeman, and J.-H. Xu, ibid., p. 1035.
7. Dr. H.-Q. Lin (private communication) has verified that the difference between sum and integral, in the typical case $a_w = a$, is as small as $(e^2/a) \times 0.007871$ for $q \to 0$.
8. The greater $a_w$, the better the integral in (16) approximates the sums in (16) or (18). The discrepancy between (5a) and (5b), in an instance where $a_w = 0$, is thus the worst we can ever expect.
9. P. Nozieres and D. Pines, Il Nuovo Cimento [X] **9**, 470 (1958).
10. Dielectric function in 2D: F. Stern, Phys. Rev. Lett. **18**, 546 (1967); P. Hawrylak, Phys. Rev. Lett. **59**, 485 (1987); P. Hawrylak, G. Eliasson, J. J. Quinn, Phys. Rev. **B36**, 10187 (1988).
11. N. Mermin and H. Wagner, Phys. Rev. Lett. **17**, 1133 & 1307 (1966).
12. S. J. Hagen, Z. Z. Wang, N. P. Ong, Phys. Rev. **B38**, 7137 (1988).
13. M. Lee et al., Phys. Rev. **B36**, 2378 (1987).

**A.J. Millis**
AT&T Bell Laboratories
600 Mountain Avenue
Murray Hill, NJ 07974

**B.G. Kotliar**
Dept. of Physics, MIT
Cambridge, MA 02139

and

**B.A. Jones**
Dept. of Physics, Harvard University
Cambridge, MA 02138

# The Two Kondo Impurity Problem: A Large *N* Biased Review

## I. INTRODUCTION

The problem of two Anderson impurities coupled to a sea of conduction electrons has attracted a lot of attention as a first step towards understanding of the interactions between local magnetic moments in a metal and the formation of the heavy electron state. In this note we would like to review briefly some aspects of this problem with an emphasis on problems which are still open. The main breakthrough in this problem was provided by the numerical Renormalization group results of Jones and Varma.[1] Our results on the large $N$ expansion appeared in Ref. (2). Other general references on the two impurities problem can be found in Ref. (7). The results of section 2 on symmetry considerations are new.

The Hamiltonian describing two Kondo impurities with a conduction electron band is given by:

$$H = H_0 + H_1 \tag{I-1a}$$

$$H_0 = -t \sum_{ij} c_{i\sigma}^\dagger c_{j\sigma} \tag{I-1b}$$

$$H_1 = J_k[S_1 \cdot \sigma(-R) + S_2 \cdot \sigma(R)]. \tag{I-1c}$$

$S_1$ and $S_2$ are the impurity spin operators located at sites $R$ and $-R$. $\sigma(R)$ is the electron spin density at site $R$. $\sigma(R) = \sum_{kk'} e^{ik-k'\cdot R} c^\dagger_{k\sigma} \sigma_{\sigma\sigma'} c_{k'\sigma'}$. $c^\dagger_{i\sigma}$ creates an electron with spin $\sigma$ at site $i$ and $c_k \equiv \frac{1}{\sqrt{N_s}} \sum_i e^{ik\cdot r_i} c_i$. $N_s$ is the number of sites. In addition to (1) it is sometimes convenient to add a direct interaction between the impurities:

$$H_d = I S_1 \cdot S_2. \quad \text{(I-1d)}$$

This problem has three energy scales. $J_R$ the *RKKY* exchange energy, which reduces to $I$ in the limit $I \gg T_k$. $T_k$ is the Kondo binding energy, which reduces to $J_k$ in the strong coupling limit $t \sim 0$. $t$ is the conduction electron bandwidth. We wish to study the ground state and low temperature properties of the model. Since the interaction is purely local it seems natural to follow Nozieres and apply a Fermi liquid approach in which the low $T$ properties are those of a Fermi gas of quasiparticles in the presence of some scattering potential. The ground state is just characterized in terms of the Fermi surface phase shifts of these quasiparticle excitations.

Two regimes can be easily identified. a) The *RKKY* regime, when $J \gg J_k$. Here the *RKKY* interaction locks the two impurities into a singlet and the conduction electrons are weakly scattered from a structureless potential. In this limit the phase shifts are close to zero. In the limit that $J \gg t \gg J_k$ this picture can be substantiated in perturbation theory in $t$ noticing that the Kondo logarithms are explicitly cut off. b) The Kondo regime when $T_k \gg J_R$. Here the Kondo effect takes place in each individual site binding two conduction electrons and leaving behind a Fermi liquid with scattering phase shifts close to $\pi/2$ at the Fermi level.

Assuming that a local Fermi liquid picture is valid there are several important issues one can address: a) How to describe quantitatively the Kondo and the *RKKY* regimes? What is the behavior of the phase shifts, the density of states close to the Fermi level and the uniform and staggered susceptibility? b) How does the transition between the two regimes take place?

## II. ROLE OF SYMMETRIES

One can often extract valuable information from symmetry considerations. In the spin 1/2 1 impurity Kondo problem, Nozieres[3] showed how many results, such as the value $\pi/2$ for the Fermi surface phase shift and the value 2 for the Wilson ratio, follow from particle-hole symmetry if a non-magnetic ground state is assumed. He and others[4] also showed how deviations from particle-hole symmetry either in the conduction band or in the Kondo coupling are irrelevant in the *RG* sense.

Here we attempt to apply similar considerations to the two impurity problem. Our new results are that if the model possesses particle-hole symmetry, then the only possible Fermi surface phase shifts are 0, $\pi/2$, or $\pi$. This result is important because it is clear on physical grounds that in the Kondo regime (for example, for widely separated impurities) the Fermi surface phase shifts must be close to $\pi/2$, while in the $RKKY$ regime the phase shifts must be close to 0 or $\pi$. It then follows that for a particle-hole symmetric model there must be a nonanalyticality between the Kondo and $RKKY$ regimes, rather than a smooth crossover. This argument does not determine the form of the nonanalyticity. As discussed in section III, the original numerical $RG$ work of Jones, Varma and Wilkins shows that in the symmetric model the nonalalyticity has the character of a second order zero temperature phase transition.

An important question, then, is whether deviations from particle-hole symmetry are relevant or irrelevant. If the symmetry-breaking perturbations are relevant, then their effect on the transition must be determined. This issue is not yet resolved. In section IV we discuss some results of a large-$N$ calculation which suggest that the particle hole symmetry breaking plays a very important role in the low energy behavior of the two impurity problem.

In the remainder of the section we define more precisely the particle-hole symmetry in which we are interested and show that it leads to the results we have stated for the phase shifts.

We are interested in $T = 0$ properties of the two impurity Kondo model. We assume that the ground state is nondegenerate and possesses a length, $L$, such that conduction electron wave packets which vanish at distances $d < L$ from the impurities evolve according to the non-interacting Hamiltonian Eq. (I-1b). In these circumstances one can learn about the phase shift by studying the conduction electron Green function, defined in imaginary time via

$$G(p, p'\tau - \tau') = -\langle T_\tau c_p(\tau) c_{p'}^\dagger(\tau') \rangle. \tag{II-1}$$

The $T$-matrix $T_{pp'}(w)$ in this problem is defined as

$$G(p, p') = G_0(p, w)\delta pp' + G(p, w)Tp, p'(w)G_0(p', w) \tag{II-2a}$$

As in any scattering problem, the phase shift $\delta_\lambda$ are defined in terms of the eigenstates of the scattering matrix $\lambda$ via the expansion of the diagonal part $t$-matrix $T_{pp}(w)$[5]

$$T_{pp}(w \rightarrow \xi_p + i\delta) = -\sum_\lambda A_\lambda^{(\xi p)} e^{i\delta_\lambda} \sin \delta_\lambda. \tag{II-2b}$$

Here $p$ is momentum, $\xi_p$ is the energy of the electron state with momentum $p$ (in the absence of the electron-impurity interaction) and $A_\lambda^{(\xi p)}$ is a real number depending on the density of states and the overlap between the plane-wave state $|p\rangle$ and the eigenstate $|\lambda\rangle$. Parity (more precisely, reflection symmetry in the plane perpendicular to and bisecting the line between the two impurities) is also a symmetry of the two impurity Kondo problem. It has been shown by Jones[6] that for fixed parity only one channel of conduction electrons couples to the two impurities. Therefore the asymptotically low energy excitations affected by the impurities are conduction electron scattering states, characterized by a phase shift which depends on energy and parity but not on spin or any other quantum number. Hence in the present case, the index $\lambda$ in II-1 runs over the values $e$ and $o$, and refers to parity only.

In problems involving electrons in metals it is sometimes convenient to work only with positive energy excitations, of which at $T = 0$ there are two sorts: adding a particle at momentum $p > p_F$ or removing a particle at momentum $p < p_F$ ($p_F$ is the Fermi momentum). Thus it is convenient to define particle operator $p_k^\dagger k \geq k_F$ and hole creation operators $h_k^\dagger$ for $k \leq p_F$ by

$$p_{k\sigma}^\dagger = c_{k\sigma}^\dagger (k > p_F)$$

$$h_{k\sigma}^\dagger = c_{-k\bar{\sigma}}(-1)^{\eta_\sigma}(k < p_F).$$

Here $\sigma$ is a spin index, $\bar{\sigma} = -\sigma$ and $\eta_\sigma = 1$ for spin up and $-1$ for spin down. One must then consider two sorts of Green functions, a particle Green function defined by

$$G_p(k, k', \tau - \tau') = -\langle T_\tau p_k(\tau) p_{k'}^\dagger(\tau') \rangle \qquad \text{(II-3)}$$

and hole Green function $G_h$, defined by

$$G_h(k, k', \tau - \tau') = -\langle T_\tau h_k(\tau) h_{k'}^\dagger(\tau') \rangle \qquad \text{(II-4)}$$

if $|k|, k' < k_F$. Using the above definitions it is easy to see that

$$G_p^R(p, p', \omega) = G^R(p, p', \omega) \; p, p' > p_F \qquad \text{(II-5a)}$$

$$G_h^R(k, k', \omega) = -G^A(-k', -k, -\omega) \; k, k' < p_F. \qquad \text{(II-5b)}$$

Similar statements hold for the self energies. The superscript $A$, $R$ denotes advanced and retarded and $\omega$ is a real frequency.

We may obtain the $T$-matrix referring to the even parity state only by studying the self energies of the even (odd) conduction electron Greens functions:

$$G_{eo}(|p|, |p'|, \tau - \tau') = -\langle T_\tau(c_p^{(\tau)} \pm c_{-p}^{(\tau)})(c_{p'}^\dagger(\tau') \pm c_{-p'}^\dagger(\tau')) \rangle. \qquad \text{(II-6)}$$

Clearly (II-5a) and (II-5b) hold separately for the even and odd Green's functions. The $\lambda = e, o$ terms in Eq. (II-2) are proportional to the diagonal part of the self energy of the even and odd Green's functions defined in Eq. (II-6).

In a theory with particle-hole symmetry the Hamiltonian must be invariant under the interchange of $p_k$ and $h_q$ operators, with $\xi_q = -\xi_k$. In the square lattice this correspondence is realized by $k = -(q + G)$ with $G = (\pi, \pi)$. Therefore

$$G_p^R(q, q', \xi_q) = G_h^R(k', k, -\xi_k) \tag{II-7}$$

Combining the statement of particle hole symmetry in Eq. (II-7) with the mathematical identity of Eq. (II-5) we have

$$G^R(q, q'\xi_q) = -G^R(-k', -k, \xi_k)^*. \tag{II-8}$$

Statement (II-8) obviously holds for the even and odd Green functions separately. The corresponding statement for the $T$ matrix gives

$$\mathbf{T}^R(|q|, \xi_q)e_0 = -\mathbf{T}^R(|k|, -\xi_k)^* e_0. \tag{II-9}$$

As $q \to p_F$, one must have $k \to p_F$, $\xi_q \to 0$, $\xi_k \to 0$ and indeed $h_k^\dagger$ and $p_q^\dagger$ must create the same state. Then the discussion following Eq. (II-2) shows that the even and odd phase shifts at the Fermi surface must be 0, $\pi/2$, or $\pi$.

We note that this entire argument is only correct asymptotically as $q \to k_F$, for it tacitly assumes that one incident particle scatters to one outgoing particle of the same energy, and not to e.g. one outgoing particle hole plus a particle pair.

We stress that the argument depends on having overall particle-hole symmetry which is a condition on the conduction electron Hamiltonian Eq. (I-1b) *and* on the exchange Hamiltonian Eq. (I-1c).

## III. THE WILSON *RG* APPROACH

In a very important development Jones and Varma and Jones, Varma, and Wilkins have obtained an exact numerical solution of a model Hamiltonian closely related to Eq. (1). Their work was the first to show that a phase transition was possible in the two impurity model. Their solution was carried out at the particle hole symmetric point. They found that the phase shifts in the Kondo phase were identically $\pi/2$, and in the *RKKY* phase they are 0 or $\pi$[8] in agreement with the general arguments developed in the previous section. The transition between the two regimes was found to be second order. As the coupling was varied through the critical value $x_c$, the specific heat coefficient and staggered susceptibility diverged approximately as $(x - x_c)^{-2}$.

## IV. THE LARGE *N* APPROACH

We have generalized the model of Eq. (I-1) to 2 $SU(N)$ impurities and solved it exactly in a large $N$ limit using the slave boson technique. This approach provides an analytic realization of the local Fermi liquid picture and allows an explicit calculation of the Fermi liquid parameters in the large $N$ limit where a mean field theory becomes exact.

As in the slave-boson treatment of the one impurity model, the mean field theory in the two impurity problem may be thought of as a resonant model. The spins are represented by localized Fermionic states which are coupled to each other by a term coming from the direct interaction Eq. (I-1d) and hybridize with the conduction electrons by a term coming from the Kondo couping, Eq. (I-1c). The physical parameters are determined by solving mean-field equations, which depend upon two dimensionless parameters. One is $T_k/J_R$ where $T_k$ is essentially the single-impurity Kondo temperature derived from $J_k$ and $J_R$ reduces to $I$ in the large $N$ limit. For $(T_k/J_R) < (T_k/J_R)_c \sim .5$, the coupling between the conduction electrons and the localized level vanishes. We interpret this as the non-Kondo regime. For $(T_k/J_R) > (T_k/J_R)_c$ the hybridization is non-zero; we interpret this as the regime in which the Kondo effect occurs. We interpret the transition between the two regimes as the large-$N$ analogue of the transition found by Jones, Varma and Wilkins.

The other parameter on which our results depend, $B$, is defined by

$$B = \left| \sum_k \frac{\cos kr}{\epsilon_k} \right| \qquad \text{(IV-1)}$$

and is seen to depend on the conduction electron bandstructure and the interimpurity separation $r$. For fixed $\epsilon_k$, $B = 0$ only at special values of $r$. When $B = 0$, the mean-field theory has particle-hole symmetry in both even and odd channels, and the Fermi surface phase shifts are found to be 0 or $\pi$ (in the non-Kondo regime) and $\pi/2$ (in the Kondo regime). When $B \neq 0$ the phase shifts differ from $\pi/2$ in the Kondo regime; however, the sum of the even channel and odd channel Fermi surface phase shifts remains $\pi$.

It is also interesting that the character of the transition between Kondo and non-Kondo regimes depends on $B$. For $B < 1/\pi$ the transition is first order; for $B > 1/\pi$ it is second order, with discontinuities only in derivatives of physical quantities such as the specific heat coefficient or the staggered susceptibility. It is interesting to speculate about the possible implications of these findings for the case $N = 2$. For the $B > 1/\pi$ regime experience with the one impurity problem suggests the transition becomes a *smooth crossover* for finite $N$. On the other hand the first order transition could in principle survive at finite $N$. There are however non perturbative effects in the form of instantons tunnelling between the Kondo and the $RKKY$ minima which would not appear in a naive $1/N$ expansion and which could potentially convert the first order transition to a second order transition or a smooth crossover. We are currently investigating these possibilities.

## V. OPEN PROBLEMS

The result of the large $N$ expansion suggest that the presence and absence of particle hole symmetry. (i.e. the parameter $B$) is a potentially relevant perturbation *at the Kondo fixed point* and *at the transition with $B = 0$ fixed point*.

It has been suggested[9] that the equality of the phase shifts in the Kondo regime is a fundamental property of the two impurity problem being an "ultimate form of confinement." If this view is correct the parameter $B$ should renormalize to zero at the Kondo fixed point.

The large $N$ approach suggests a very different scenario. Here the phase shifts are generically unequal in the absence of particle hole symmetry. Their splitting is the first signature of the formation of a heavy Fermion band in the two impurity problem. We note that in the large $N$ calculation the effect of $B$ on the phase shifts appears at very low energies (of the order of $Tk$) indicating that the splitting of the resonances is also a low energy effect, which would then not renormalize to zero, unlike particle-hole asymmetry in the 1-impurity problem.[4]

These two different scenarios could be clarified by a numerical $RG$ in the Kondo phase of a two impurity problem lacking particle hole symmetry. If $B$ is indeed not driven to zero at low energies it is important to understand whether it could have an effect on the nature of the phase transition.

If the solution of the two impurity problem is to yield some clue about the lattice problem, it will be by understanding how the resonance splitting is renormalized by many body effects. We are only beginning to address this point. Expansions about solvable limits will provide some further insight, but at present it seems to us that only a numerical $RG$ study or an exact solution of a particle hole asymmetric model will definitively settle this issue.

## VI. REFERENCES

1. B. A. Jones and C. M. Varma, *Phys. Rev. Lett.* **58**, 843 (1987) and B. A. Jones, C. M. Varma and J. W. Wilkins, *Phys. Rev. Lett.* **61**, 125 (1988).
2. B. A. Jones, B. G. Kotliar, and A. Millis, *Phys. Rev. B* (1988) in press.
3. P. Nozieres and A. Blandin, *J. Physique* **41**, 193 (1980).
4. P. Lloyd and D. M. Cragg, *J. Phys. C* **12**, 3290 (1979); D. M. Cragg and P. Lloyd, *J. Phys. C* **11**, L597 (1978) and *J. Phys. C* **12**, L215 (1979); D. M. Cragg, P. Lloyd, and P. Nozieres, *J. Phys. C* **13**, 803 (1980).
5. G. D. Mahan, *Many Particle Physics*, section 4.1 (Plenum Press, New York) 1983.
6. B. A. Jones, unpublished Ph.D. Thesis, Cornell University, 1987.
7. P. Coleman, *Phys. Rev. B* **35**, 5072 (1987); C. Jayaprakash, H. R. Krishnamurthy, and J. W. Wilkins, *Phys. Rev. Lett.* **47**, 737 (1981); H. Ishii, *Progress of Theoretical Physics*, **50**, 1777 (1973).

8. B. A. Jones in these proceedings.
9. C. M. Varma and B. A. Jones, "Theoretical Effects of Valence Fluctuations and Heavy Fermions," L. C. Gupta and S. K. Malik editors, Plenum, NY, 878 (1987); C. M. Varma, *Phys. Rev. Lett.* **55**, 2723 (1985).

**M. Rasolt**
Solid State Division
Oak Ridge National Laboratory
Oak Ridge, Tennessee 37831

Lyman Laboratory of Physics
Harvard University
Cambridge, MA 02138

and

**Z. Tesanovic**
Department of Physics and Astronomy
The Johns Hopkins University
Baltimore, MD 21218

# A New Crossover Behavior of Type II Superconductors in Strong Magnetic Fields

An extended magnetic field *does not* destroy superconductivity in type II superconductors. As the field increases, the Abrikosov flux lattice crosses over into a *new quantum limit*, characterized by a transition temperature which is an *increasing* function of the field, the absence of the Meissner effect and a supercurrent flow along the field direction. The transition temperature remains finite in an *arbitrarily* strong external field.

We consider, in this section, a many electron system in the presence of a uniform external magnetic field $\vec{H}$, with some mechanism for attractive interparticle interaction. We focus exclusively on the effect of a uniform magnetic field on the superconducting condensate over the *full* range of $\vec{H}$. We show that as $\vec{H}$ increases the condensate crosses over *continuously* from a type II superconductor to a remarkable new superconducting state whose High field limit (Quantum limit) was discussed recently by Rasolt[1] and Tesanovic and Rasolt.[2]

The classical theory of pure type II superconductivity leads to a critical temperature which decreases to zero as a function of the critical field $H_{C2}$ (see Fig. 1). $T_c$ is *identically* zero when $H_{C2}(T=0) = \phi_0/(2\pi\xi_0^2)$, where $\phi_0$ is the elementary flux and $\xi_0$ is the coherence length at zero temperature. In fact, the *classical* result is *not true!* The intrinsic Quantum make up of the order parameter

**167**

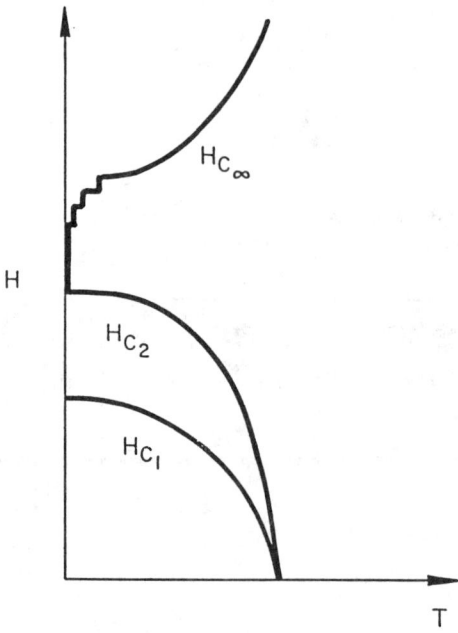

FIGURE 1 A sketch of the phase diagram for the superconducting condensate in the $T - H$ plane. As the field increases one crosses over from a Meissner state to the one-dimensional high quantum limit state at $H_{c\infty}$. The step-like behaviour in the cross-over region reflects the population of higher Landau levels.

(*i.e.*, the fact that it is made up of *discrete* Landau levels; *no matter how many* of them) will never permit $T_c$ to go *to zero*. In this low temperature region of $H_{C2}$ we find a novel crossover to a totally new superconducting state (along $H_{C\infty}$) where $T_c$ is strongly enhanced as $\vec{H}$ increases (see Fig. 1). We also find that this crossover evolves into a new relation between the magnetic field and the order parameter, which in the Quantum limit is entirely different from the Landau relation; appropriate in the "classical" region. Finally we consider the role of fluctuations, impurities and the Zeeman splitting on this new superconducting condensate.

The partition function $Z$ for a weak coupling superconductor in the presence of an arbitrarily strong uniform magnetic field $\vec{H}$ can be written in a path integral form as

$$Z = \int D\phi_\sigma(\vec{r}) D\phi_\sigma^*(\vec{r}) D\vec{a}(\vec{r}) e^{-\int_0^\beta d\tau L(\tau)} \quad (1a)$$

where

$$L(\tau) = \sum_{\vec{r},\sigma} \left\{ \phi_\sigma^*(\vec{r}) \left[ \frac{\partial}{\partial \tau} + \frac{1}{2m} \left( \hbar \frac{\partial}{\partial \vec{r}} - \frac{ie}{c}(\vec{A}(\vec{r}) + \vec{a}(\vec{r})) \right)^2 \right] \phi_\sigma(\vec{r}) \right. \tag{1b}$$
$$\left. - \frac{\gamma}{2} \phi_\sigma^*(\vec{r})\phi_\sigma(\vec{r})\phi_{-\sigma}(\vec{r})\phi_{-\sigma}(\vec{r}) - \frac{g\hbar e}{2mc} H - \frac{|\vec{\nabla} \times (\vec{A}(\vec{r}) + \vec{a}(\vec{r}))|^2}{8\pi} \right\}$$

In Eq. 1a $\vec{A}(\vec{r})$ is the vector potential corresponding to the uniform part of the magnetic field;[5] we choose to work in the symmetrical Gauge so $\vec{A}(\vec{r}) = \frac{1}{2}\vec{H} \times \vec{r}$. The $D$ is shorthand for $\Pi_{\vec{r},\sigma} D\phi_\sigma(\vec{r}) \ldots$ etc. and $\phi_\sigma(\vec{r})$, $\phi_\sigma^*(\vec{r})$ are anticommuting Grossman number. In Eq. 1b we have replaced the dynamic and spatially nonlocal interparticle interaction with static point interaction $+\gamma\delta(\vec{r}_1 - \vec{r}_2)$. (An approximation (generally applied in the weak coupling limit). We also cut off this interaction outside the energy range $E_0$ around the Fermi energy $\mu$. The above static approximation is irrelevant to the interest here. The choice of a point interaction will become important (particularly in the high Quantum limit) when we discuss the role of fluctuations (see below). Finally, we have also added an internally induced field $\vec{a}(\vec{r})$ in the expectation of broken symmetry, below $T_c$, whose order parameter generates a self-consistent field $\vec{a}(\vec{r})$. We can now calculate the following three thermodynamic quantities: $\langle \phi_\sigma^*(\vec{r})\phi_{-\sigma}^*(\vec{r}) \rangle$, $\langle \phi_\sigma^*(\vec{r})\phi_\sigma(\vec{r}) \rangle$ and $\langle \vec{a}(\vec{r}) \rangle$ (where the brackets are the functional integral over the Fermion fields $\phi^*$, $\phi$ and electromagnetic field $\vec{a}(\vec{r})$[6] in the partition function of Eq. 1, divided by $Z$). To get the set of self-consistent equations displayed in Fig. 2; to first order in the coupling $\gamma$.

To calculate $T_c$ we need not consider the tadpole equation of Fig. 2c, it will enter only in the structure of the condensate below $T_c$ (see below). Fig. 2 then leads to the following integral equation for $T_c$

$$\Delta(\vec{r}, Z) + \gamma \int dZ' dx' dy' K_{\sigma,-\sigma}(\vec{r}, \vec{r}', Z - Z') \Delta(\vec{r}', Z') = 0 \tag{2a}$$

where $\vec{r} = (x, y)$ and $Z$ is along $\vec{H}$. In Eq. 2a

$$K_{\sigma,-\sigma}(\vec{r}, \vec{r}', Z - Z') = \sum_{\substack{n,n' \\ k_z}} K_{n,n' \atop \sigma,-\sigma}(\vec{r}, \vec{r}', Z - Z') \tag{2b}$$

$$K_{n,n' \atop \sigma,-\sigma}(\vec{r}, \vec{r}', Z - Z') =$$

$$+ \beta^{-1} \sum_{\xi_\ell} G_{\sigma,n'}(\xi_\ell, \vec{r}, \vec{r}', Z - Z') G_{-\sigma,n}(-\xi_\ell, \vec{r}, \vec{r}', Z - Z') \tag{2c}$$

(a)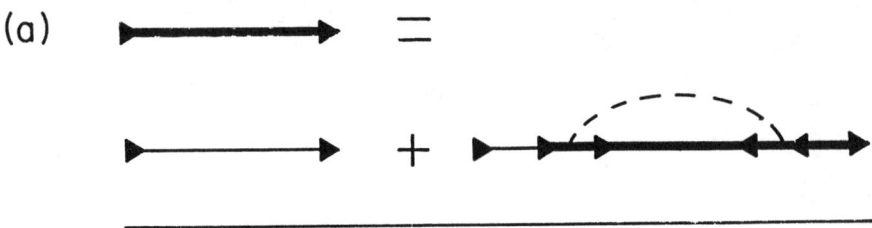

(b)

(c)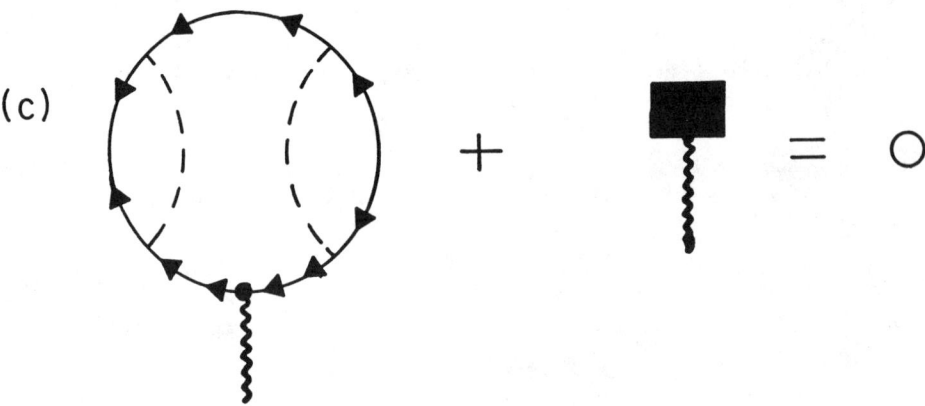

FIGURE 2 Feynman graphs representing the self-consistent equations for $\langle \phi_\sigma^*(\vec{r})\phi_\sigma(\vec{r})\rangle$, $\langle \phi_\sigma^*(\vec{r})\phi_{-\sigma}^*(\vec{r})\rangle$ and $\langle \vec{a}(\vec{r})\rangle$ to lowest order in $\gamma$. (a) The equation for the normal interacting propagator $\langle \phi_\sigma^* \phi_\sigma \rangle$ given by the doubly solid line terminating with two arrows pointing in the same direction. The single solid line is the noninteracting normal propagator, the doubly solid line terminating with two arrows pointing in opposite direction is the anomalous propagation $\langle \phi_\sigma^* \phi_{-\sigma}^* \rangle$ and the dashed line is the interparticle interaction $\gamma\delta(\vec{r}_1 - \vec{r}_2)$. (b) The equation for the anomalous propagator $\langle \phi_\sigma^* \phi_{-\sigma}^* \rangle$. (c) the tadpole equation for the self-consistently induced $\langle \vec{a}(\vec{r})\rangle$. In the black box the tadpole couples to the electromagnetic energy. The other term couples $\vec{a}$ to the electrons and the coupling vertex (the black circle) is $\vec{a} \cdot \left(\hbar \frac{\partial}{\partial \vec{r}} - \frac{ie}{c}\vec{A}(\vec{r})\right) + \left(\hbar \frac{\partial}{\partial \vec{r}} - \frac{ie}{c}\vec{A}(\vec{r})\right) \cdot \vec{a}$.

# A New Crossover Behavior of Type II Superconductors

$$G_{\sigma,n}(\xi_\ell, \vec{r}, \vec{r}', Z - Z') = \sum_{k_z,m} \frac{e^{ik_z(Z-Z')}}{\sqrt{L}} \frac{\phi^*_{n,m}(z)\phi_{n,m}(z')}{\xi_\ell - \frac{\hbar^2 k_z^2}{2m} - \hbar\omega_n + \mu} \quad (2d)$$

and where $z = x + iy$ and

$$\phi_{n,m}(z) = (2^{m+n})^{1/2}/(2\pi m!n!)^{1/2} \exp\left(\frac{1}{4}|z|^2\right)\left(\frac{\partial}{\partial z^*}\right)^m \left(\frac{\partial}{\partial z}\right)^n \exp\left(-\frac{1}{2}|z|^2\right) \quad (2e)$$

In Eq. 2 $k_z$ is the momentum parallel to $\vec{H}$, $L$ is the macroscopic length of the system along $\vec{H}$, $\xi_\ell = \frac{\pi i}{\beta}(2\ell + 1)$ and $\omega_n = neH/mc$. The chemical potential is a function of $H$ and the electron density $\rho$ i.e., $\mu = \frac{\hbar^2 k_F^2}{2m}$ where

$$\sum_{n \text{ occupied}} \left(k_F^2 - \frac{2m}{\hbar}\omega_n\right)^{1/2} = \pi^2 \ell_0^2 \rho$$

where $\ell_0 \equiv \left(\frac{eH}{\hbar c}\right)^{-1/2}$. We have ignored in Eq. 2 the Zeeman splitting[8] (i.e., we set $g = 0$); we discuss the effect of a finite $g$ shortly. Eq. 2 is general enough to incorporate *any size* of $\vec{H}$ provided we choose $E_0 \ll \frac{\hbar^2 k_F^2}{2m}$ (i.e., weak coupling) which we can always do. The solution of Eq. 2 must then describe the crossover from $H_{C2}$ all the way to $H_{C\infty}$ (see Fig. 1). The solution of Eq. 2 for $T_c$ is a very nontrivial matter. It hinges on the observation that *any* function $f(z)$, made up from the basis set of Eq. 2e in the *lowest* Landau level $n = 0$, satisfies the following relation

$$\int dx' dy' K_{n,n'}^{\sigma,-\sigma}(\vec{r}, \vec{r}', 0)f(z') = A_{n,n'}f(z) \quad (3a)$$

where $A_{n,n'}$ are complicated functions of $H$, $T$ and $\rho$ and will not be detailed here.[9] Equation (3) follows a lengthy calculation which will also be presented elsewhere.[9] We next write the order parameter $\Delta(\vec{r}, Z)$ as

$$\Delta(\vec{r}, Z) = f(z)\left[\sum_q e^{iqZ}\Delta(q)\right]. \quad (4)$$

In the weak coupling limit two different regions of $q$ are important. $q = 0$ provides for a uniform variation along $\vec{H}$. However, a nonuniform component along $\vec{H}$

cannot be ruled out. From Eq. 2 it is not difficult to see that such nonuniform singular contributions arise when

$$q \equiv q_{n,n'} = \left(k_f^2 - \frac{2m}{\hbar}\omega_n\right)^{1/2} - \left(k_f^2 - \frac{2m}{\hbar}\omega'_n\right)^{1/2}. \tag{5}$$

Using then Eqs. (3) and (4) in Eq. 2a we first find that the uniform solution in Eq. (4) (i.e., $q = 0$) always has the higher $T_c$. We have solved these Equations for $T_c$ in this uniform state ($q = 0$) up to 30 Landau levels (see Fig. 1) and observed a continuous oscillatory decrease in $T_c$ for lower fields or equivalently higher number of occupied Landau levels $n_c$ (we are needing ρ constant). In fact, in this region to sufficient accuracy, we can largely neglect the nonsingular contributions (nonsingular in $T$) coming from the off-diagonal terms ($n \neq n'$) in Eq. (3) to get a closed form (always in weak coupling) result for $T_c$ given by

$$T_c \approx 1.14\Omega\exp\left\{-\frac{2\pi\ell_0^2}{V}\left(\sum_{n=0}^{n_0} N_{1n}(0)t(n)\right)^{-2}\right\}, \tag{6}$$

where $N_{1n}(0)$ is the one-dimensional density of states for the $n$-th Landau level and

$$t(n) = (n!)^2 \sum_{i,j=0}^{n} \frac{(-)^{i+j}(i+j)!}{2^{i+j}(i!j!)^2(n-i)!(n-j)!} \equiv \frac{(2n)!}{2^{2n}(n!)^2}$$

For a much larger $n_c$ a numerical solution becomes much more difficult. However by carefully examining Eqs. (2a), (3) and (4) we find that $T_c$ *can never go to zero*. As we reach the crossover to $H_{C2}$ we are crossing over from the energy scales

$$\hbar\omega_c \gtrsim \frac{\hbar^2}{2m}k_f^2 \gg E_0 \gg k_B T_c$$

to

$$\frac{\hbar^2 k_f^2}{2m} \gg E_0 \gg k_B T_c \gg \hbar\omega_c.$$

In this region these off-diagonal (i.e., $n \neq n'$) nonsingular (in $T$) contributions grow rapidly leading to an increase in $T_c$ along $H_{C2}$; as $H$ decreases. In this region we can show that Eq. 2 crosses over to the usual Landau-Ginsburg (LG) form.[10] Of course the LG form leads to a $T_c = 0$ when $H_{C2} = \phi_0/(2\pi\xi_0^2)$. But when $T_c \to 0$ Eq. 2 crosses over back to the region where $\hbar\omega_c \gg k_B T$ and where the LG form is no longer valid.[10,11] In short, we discover that a rigorous quantum mechanical

# A New Crossover Behavior of Type II Superconductors

treatment of a weak coupling pure type II superconductor (neglecting Pauli breaking) leads to a *finite $T_c$ for any size of the magnetic field*.

In our discussion, so far, we only considered the behavior of $T_c$. There remains however, a serious conceptual problem: How does the increase in $T_c$ along $H_{C\infty}$ square with the Meissner effect? The answer depends on the nature of the superconducting order parameter and its relation to the induced vector potential $\vec{a}(\vec{r})$. We study this relation close to the critical line.

Close to the critical line (see Fig. 1) we can iterate the self-consistent equations (displayed in Fig. 2) to third order in $\Delta(\vec{r}, Z)$ (and $\Delta^*(\vec{r}, Z)$) and linear order in $\vec{a}(\vec{r})$ to get two coupled equations between $\Delta(\vec{r}, Z)$ and $\vec{a}(\vec{r})$. To analyze these two equations over the full range of $\vec{H}$ is complicated. We can however, get important insight to the full solution by studying the Quantum limit, where all the electrons occupy the lowest Landau level, (*i.e.*, the region above the last oscillation in Fig. 1). In this range of the critical line, and again considering only the weak coupling limit, these two coupled equations decouple to the following two relations:

For the order parameter we get

$$\alpha(T) \int d^2r_2 (K_2(\mathbf{r}_1, \mathbf{r}_2)\Delta(\mathbf{r}_2))$$

$$+ \beta(T) \int d^2r_2 d^2r_3 d^2r_4 \Delta(\mathbf{r}_2) K_4(\mathbf{r}_1, \mathbf{r}_2, \mathbf{r}_3, \mathbf{r}_4) \Delta^*(\mathbf{r}_3)\Delta(\mathbf{r}_4) = 0. \quad (7a)$$

For the induced $\vec{a}(\vec{r})$, we get

$$\frac{\partial b}{\partial z}(z, z^*) = \frac{-\gamma^2 e^2 H}{\pi \ell_0^2 mc^2} C_2 z^* e^{-|z|^2/2} \int d^2r_1 \int d^2r_2 \Delta^*(\vec{r}_1) \Delta(\vec{r}_2)$$

$$\times e^{\left(-\frac{|z_1|^2}{2} - \frac{|z_2|^2}{2} + \frac{z_1 z_2^*}{2} + \frac{z_1 z^*}{2} + \frac{z z_2^*}{2}\right)} \quad (7b)$$

and in Eq. 7a

$$K_2(\mathbf{r}_1, \mathbf{r}_2) = \frac{1}{(2\pi\ell_0^2)^2} \exp(-z_1^* z_1/2 - z_2^* z_2/2 + z_1 z_2^*), \quad (7c)$$

$$K_4(\mathbf{r}_1, \mathbf{r}_2, \mathbf{r}_3, \mathbf{r}_4) = \frac{1}{(2\pi\ell_0^2)^4} \exp(-z_1^* z_1/2 - z_2^* z_2/2 - z_3 z_3^*/2$$

$$- z_4^* z_4/2 + (z_1 + z_3)(z_2^* + z_4^*)/2). \quad (7d)$$

$$\alpha(\tau) = \gamma + \frac{\gamma^2}{\beta} \sum_{k_z, \xi_\ell} \frac{1}{[\xi_\ell - \epsilon(k_z)][\xi_\ell + \epsilon(k_z)]} \quad (7e)$$

$$\beta(\tau) = \frac{\gamma^2}{\beta} \sum_{k_z \xi_\ell} \frac{1}{(\xi_\ell - \epsilon(k_z))^2} \frac{1}{(\xi_\ell + \epsilon(k_z))^2} \qquad (7f)$$

$$C_2 = \sum_{\xi_\ell, k_z} \frac{\epsilon(k_z)}{(|\xi_\ell|^2 + \epsilon(k_z)^2)^2} \qquad (7g)$$

with $\vec{b}(z, z^*) = \vec{\nabla} \times \vec{a}(\vec{r})$. Finally, $\epsilon(k_z) = \dfrac{\hbar^2 k_z^2}{2m} - \mu$. The derivation of Eq. 7 is somewhat lengthy and will not be presented here.[9] The crucial point is that the relation between $\vec{a}(\vec{r})$ and the order parameter $\Delta(\vec{r})$ (Eq. 7b) is entirely different from the Landau relation (or further modifications[12]). In fact we can solve Eq. 7 variationally.[9] The solution is the same as the Abrikosov[3] hexagonal lattice *however*, with the penetration depth replaced by the *magnetic length* $\ell_0^* = \ell_0/\sqrt{2}$. We also find from Eq. 7b that the local field $\vec{b}(z, z^*)$ peaks where the superconducting density peaks unlike in the classical limit. In short in the Quantum limit the field is very strong and penetrates everywhere into the superconductor; indeed the superconductivity is due to such a strong field. Consequently the relationship between the supercurrent and the induced vector potential $\vec{a}(\vec{r})$ is entirely different from that given by the Landau equation. As we increase the number of Landau levels the direct solution of Eq. 7 becomes more complicated. However, in view of our discussion for $T_c$ (which is really the solution of 7a with all Landau levels included without however, the quartic term) we expect that for increasing number of Landau levels the system will continuously cross over to the usual LG form (and therefore to the London relation) along $H_{C2}$.[13]

Eqs. 2 and 3 (for small $n$, $n'$) look very one-dimensional. In fact, in the quantum limit (*i.e.*, $n = n' = 0$) they are strictly one-dimensional. This then brings the important question of fluctuations. Unfortunately we have no rigorous results for the effect of such quantum fluctuations on the critical line.[14] However, we can explicitly demonstrate that infact the system is *not truly one-dimensional*. In one-dimensional, the weak coupling set of ladder graphs, which evolved from Fig. 2a, are not sufficient. The point is that both electron-electron and electron-hole graphs lead to $\ln(k_B T/\mu)$ behavior and therefore must be treated consistently.[15] In fact, even the point interaction is insufficient and must be replaced by two values; a forward cross-section $\gamma = g_1$ and a backward cross-section $\gamma = g_2$ (see Fig. 3a). The set of graphs in Fig. 3b are extension of our weak coupling with now two different cross sections; these will continue to be one-dimensional. However, in Fig. 3c the cross-graphs *cannot* be reduced to an one-dimensional form; the intermediate sums over the degeneracy in the Landau levels will not permit this. Such contributions in a true one-dimensional situation are crucial to the elimination of long range order (LRO). In our case such terms will be different and LRO can exist at finite $T$.[16]

We have so far neglected the effect of impurities and Zeeman splitting. Impurities will add an additional energy scale $\hbar/\tau$. The scattering will mix the various Landau levels and will reduce (or eliminate) the range of the continuous

# A New Crossover Behavior of Type II Superconductors

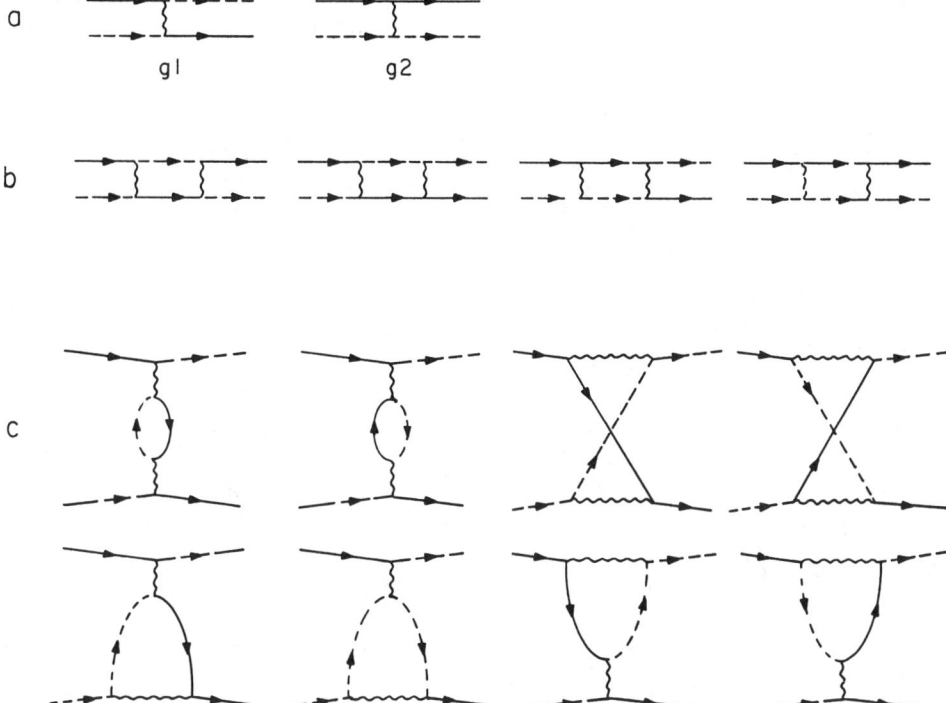

FIGURE 3 Low-order scattering processes (to second order) in γ in one-dimensional Fermi systems (see Ref. 5). (a) $g_1$ is a backward scattering process and $g_2$ is a forward process. (b) Some ladder scattering contribution standard for the weak coupling limit. (c) Non-mean-field contributions, some of which no longer reduce to a strictly one-dimensional form, when the scattering electrons are placed in a very strong magnetic field; i.e. lowest Landau level.

crossover region when $\hbar/\tau \gg \hbar\omega_c \gg k_B T_c$.[17] The superconductors with low $T_c$ (and correspondingly small $H_{C2}$) impurities will make the cross over region. For high $T_c$ materials however, this is not the case (of course very high magnetic fields are required).[18]

For a finite $g$, in the quantum limit, the spin singlet will eventually rupture. Here we can turn to the exciting possibilities in multi-valley semiconductors (i.e. isospin space where $g = 0$ discussed in ref. (1)).[19] In the intermediate field region the spin singlet will survive inspite of the Zeeman splitting. In fact, the Zeeman splitting will contribute a modulation along $\vec{H}$ with

$$q_{\sigma,-\sigma} = \left(k_f^2 - \frac{2m}{\hbar}\left(\omega_n + \frac{geH}{2mc}\right)\right)^{1/2} - \left(k_f^2 - \frac{2m}{\hbar}\omega_n\right)^{1/2}$$

(see $q_{n,n'}$ in the context of Eq. 5) which will significantly reduce the Pauli pair-breaking.

In conclusion we have solved for the critical line of a weak coupling type II superconductor over the full range of the magnetic field and have shown a crossover from the classical $H_{C2}$ line to a remarkable superconducting state. The implication of our results extend beyond condensed matter physics to any areas where the quantum make up of the Abrikosov lattice plays an important role.

## ACKNOWLEDGEMENTS

This work has been supported in part by the National Science Foundation through Grant No. DMR85-14638 and the Harvard Materials Research Laboratory, and in part by the Division of Materials Sciences, U.S. Department of Energy under Contract No. DE-AC05-84OR21400 with Martin Marietta Energy System. Z. T. acknowledges the support of the Packard Foundation and the hospitality of the Aspen Center for Physics where part of the work was performed.

## REFERENCES

1. M. Rasolt, *Phys. Rev. Lett.* **58**, 1482 (1987).
2. Z. Tesanovic and M. Rasolt, *Phys. Rev.* **B39**, 2718 (1989).
3. A. A. Abrikosov, *Zh. Eksf. Teor Fcz* **32**, 1442 (1957).
4. W. A. Kleiner, L. M. Roth and E. H. Autler, *Phys. Rev.* **A133**, 1226 (1964).
5. It also includes (below $T_c$) the uniform part of the induced magnetic field.
6. Note that these fluctuations are around the induced $\tilde{a}(\vec{r})$ and are therefore finite $\tilde{q}$ fluctuations which are not relevant to the superconducting transition.[7]
7. B. I. Halperin, T. C. Lubensky and S. K. Ma, *Phys. Rev. Lett.* **32**, 292 (1974).
8. Systems where $g = 0$ do exist; see M. Rasolt, B. I. Halperin and D. Vanderbilt, *Phys. Rev. Lett.* **57**, 126 (1986).
9. To be published.
10. P. G. de Gennes, "Superconductivity of Metals and Alloys," Eq. 6-63. See also discussion in Chapter 7.
11. The point is that the continuous form of the LG equation is achieved only if the oscilations in $\hbar\omega_n$ are smoothed out. This occurs only when $k_B T \gg \hbar\omega_c$. A useful analog occurs in calculating the Landau noninteracting diamagnetic susceptibility at $T = 0$, where the proper limits are again $k_B T \to \hbar\omega_c \to 0$.
12. A. B. Pippard, *Proc. Roy. Soc.* (London) **A216**, 547 (1953).
13. We expect the Abrikosov solution to continue to be valid along $H_{C\infty} \to H_{C2}$. Of course, along $H_{C2}$ the relation of $\ell_0^*$ to $T$ is not given by $\ell_0^* =$

$$\sqrt{\pi}\xi_0\left(1 - \left(\frac{T}{T_c}\right)^2\right)^{-1/2}.$$

14. Long wavelength classical fluctuations along $\vec{H}$ have been studied below six dimensions by E. Brezin, D. R. Nelson and A. Thiaville, *Phys. Rev.* **B31**, 7124 (1985). They find the transition to be driven weakly first order. Their results are of course, also relevant here.
15. J. Sölyom, *Advances in Phys.* **28**, 201 (1979).
16. Of course, the precise relation of $T_c$ vs $H$ (*i.e.*, Eq. 6) will be significantly modified; particularly in the Quantum limit.
17. It has been already recognized in the classical region (see Ref. 10) that impurities will increase the range of validity of the LG form.
18. In $La_2CuO_4$ we estimate the field to be in the mega-Gauss range.
19. Incidentally, it can be shown that a magnetic field will generally reduce the repulsive $e - e$ interaction in $\gamma$ and increase the attractive component as $H$ increases[1] thus favoring more the superconducting instability over many other types of instabilities available in the quantum limit.

**N. Read and Subir Sachdev**
Center for Theoretical Physics and Section of Applied Physics
P.O. Box 6666 and P.O. Box 2157
Yale University
New Haven, CT 06511

# Valence-Bond and Spin-Peierls Ground States of Low-Dimensional Quantum Antiferromagnets

The large-$N$ limit of a nearest-neighbor $SU(N)$ antiferromagnet on a bipartite lattice exhibits in dimensions $d \geq 2$ a zero-temperature phase transition between a Néel-ordered state and a resonating-valence-bond state. Here it is shown in $d = 1, 2$ that topological effects produce spin-Peierls or valence-bond-solid order in the non-Néel phase with a ground-state degeneracy which varies periodically with "spin" for fixed $N$, with periodicity given by the coordination number of the lattice. Thus a non-Néel phase of the spin-$\frac{1}{2}$ Heisenberg model on a square lattice would be a spin-Peierls state with a fourfold degeneracy due to broken lattice rotational symmetry.

Following the discovery of high-temperature superconductivity,[1] it has been proposed that the phenomenon is linked to a $T = 0$ disordered (i.e., non-Néel) phase of the Heisenberg antiferromagnet on a square lattice.[2] We examine here nearest-neighbor generalizations of the standard Heisenberg model on bipartite lattices in dimensions $d = 1, 2$ all of which have a two-sublattice Néel state as their classical ground state. We find that *topological effects radically influence the nature of the disordered phase, producing in general a spin-Peierls or valence-bond-solid state.* The degeneracy of this state varies periodically with the mag-

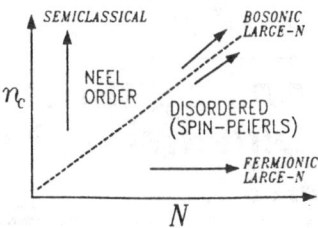

FIGURE 1 Phase diagram of the square-lattice $SU(N)$ antiferromagnet as a function of the spin $n_c$ [$=2S$ for $SU(2)$]. The phase boundary between Néel order and its absence behaves as $n_c/N \to 0.19$ as $N \to \infty$ (Ref. 6). Earlier work examined the semiclassical (Refs. 3 and 4) and the fermionic large-$N$ limits (Refs. 4, 6, and 7); the latter has spin-Peierls order with the symmetry of Fig. 2(d) for *all* $n_c$. This paper examines the bosonic large-$N$ region in the disordered phase close to the transition line. In $d = 1$, the Néel region is absent, while for $d > 2$, a similar phase boundary is found (Ref. 4).

nitude of the "spin" at each lattice site in accordance with the recent prediction of Haldane[3] and its generalization of $SU(N)$.[4] In $d = 2$, the elementary spin excitations are confined (i.e., permanently bound) pairs of "spinons" and there is a spinless collective mode with an energy gap at all wave vectors.[5] Our results for the phase diagram and ground states are summarized in Figs. 1 and 2.

We study a family of models with Hamiltonian

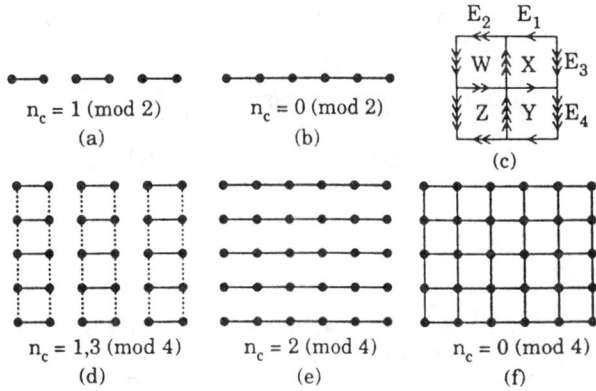

FIGURE 2 Symmetry of the ground states: Solid lines denote larger values of $\langle \hat{S}(i) \cdot \hat{S}(i+1) \rangle$ for a link; no line, smaller values; and dashed line, intermediate values. (a), (b) $d = 1$ chain. (c) Definition of the four plaquette sublattices $W$, $X$, $Y$, $Z$ and the electric fields on the links. (d)–(f) Symmetry of ground states for square lattice near phase boundary in Fig. 1, of degeneracy 4, 2, 1, respectively.

## Valence-Bond and Spin-Peierls Ground States

$$H = \frac{J}{N} \sum_{\langle i,j \rangle} \hat{S}^\beta_\alpha(i) \hat{S}^\alpha_\beta(j), \quad (1)$$

where $\hat{S}^\beta_\alpha(i)$ are the generators of $SU(N)$, $\langle i, j \rangle$ denotes pairs of nearest neighbors ("links") on a $d$-dimensional hypercubic lattice, and repeated indices $\alpha, \beta = 1, \ldots, N$ are summed over. We will use a Schwinger boson representation of the spin states, in which $\hat{S}^\beta_\alpha(i) = b^\dagger_\alpha(i) b^\beta(i)$, $i \in A$ sublattice, and $\hat{S}^\beta_\alpha(j) = -\bar{b}^{\beta\dagger}(j)$, $j \in B$ sublattice; the $\bar{b}$ bosons are implies by the placement of indices to transform as the conjugate representation to $b$, which are in the fundamental representation of $SU(N)$. If we impose the constraint $b^\dagger_\alpha b^\alpha = n_c$ or $\bar{b}^{\alpha\dagger} \bar{b}_\alpha = n_c$ at each site, then the states are in an irreducible representation of $SU(N)$ which has a Young tableau with one row of $n_c$ boxes (the totally symmetric representation) on sublattice $A$, and the conjugate of this on sublattice $B$ ($N - 1$ rows, $n_c$ columns). In the familiar case of $N = 2$ [$SU(2)$ or $O(3)$ Heisenberg model], these are the usual Schwinger bosons, and all sites have spin $S = n_c/2$. This representation has been used previously by Arovas and Auerbach[6] to obtain a $1/N$ expansion with $n_c \propto N$ in order to study mainly the Néel-ordered phase.

We may represent the partition function of our models by an imaginary-time functional integral[8] of

$$\mathcal{L} = \sum_{i \in A} \left[ b^\dagger_\alpha(i) \left( \frac{d}{d\tau} + i\lambda(i) \right) b^\alpha(i) - i\lambda(i) n_c \right] + \sum_{j \in B} \left[ \bar{b}^{\alpha\dagger}(j) \left( \frac{d}{d\tau} + i\lambda(j) \right) \bar{b}_\alpha(j) \right.$$

$$\left. - i\lambda(j) n_c \right] + \sum_{i \in A, \hat{\eta}} \left[ \frac{N}{J} |Q_{i,i+\hat{\eta}}|^2 - Q^*_{i,i+\hat{\eta}} b^\alpha(i) \bar{b}_\alpha(i + \hat{\eta}) + \text{H.c.} \right)$$

over the fields $b$, $\bar{b}$, $Q$, and $\lambda$. Here the $\lambda(i)$ fix the boson number at each site, $\tau$ dependence of all fields is implicit, $Q$ was introduced by a Hubbard-Stratonvich decoupling of $H$, and $\hat{\eta}$ runs over nearest-neighbor vectors and has length $a$. The Lagrangian $\mathcal{L}$ possesses a $U(1)$ gauge invariance under arbitrary $\tau$-dependent changes of phase of $b$, $\bar{b}$, provided corresponding changes in $Q$, $\lambda$ are made; the functional integral over $\mathcal{L}$ faithfully represents the partition function as long as we fix a gauge, e.g., by the condition $d\lambda/d\tau = 0$ at all sites.

The $1/N$ expansion of the free energy can be obtained by integrating out of $\mathcal{L}$ the $N$-component $b$, $\bar{b}$ fields to leave an effective action for $Q$, $\lambda$ having coefficient $N$ (since $n_c \propto N$); minimizing with respect to the "mean-field" values of $Q$, $\lambda$ gives the $N \to \infty$ limit.[8] This is equivalent to solving the mean-field Hamiltonian

$$H_{\text{MF}} = \sum_{i \in A, \eta} [N|\bar{Q}|^2/J - \bar{Q} b^\alpha(i) \bar{b}_\alpha(i + \hat{\eta}) + \text{H.c.}] + \bar{\lambda} \sum_{i \in A} [b^\dagger_\alpha(i) b^\alpha(i)$$

$$- n_c] + \bar{\lambda} \sum_{j \in B} [\bar{b}^{\alpha\dagger}(i) \bar{b}_\alpha(i) - n_c].$$

In writing $H_{\text{MF}}$ we used the fact that $i\lambda(i) = \bar{\lambda}$ and $Q_{i,i+\eta}$ are found to be uniform and independent of $\hat{\eta}$ at the saddle point. The constant $\bar{\lambda}$ is found to be real and

$\bar{Q}$ can be taken real, positive by a gauge transformation. The Hamiltonian $H_{\rm MF}$ can be diagonalized by Bogoluibov's method and we find two modes for each wave vector in the (reduced) Brillouin zone, of energy $\omega_k = (\bar{\lambda}^2 - 4d^2\bar{Q}^2\gamma_k^2)^{1/2}$, where $\gamma_k = (1/2d)\Sigma\eta e^{ik\cdot\eta}$ and $\bar{\lambda} \sim \bar{Q} \sim J$. At $\mathbf{k} = 0$, $\omega_k = \Delta = (\bar{\lambda}^2 - 4d^2\bar{Q}^2)^{1/2} \geq 0$ is the energy gap. In $d = 1$, $\Delta \to 0$ as $n_c/N \to \infty$; in $d = 2$, $\Delta \to 0$ as temperature $T \to 0$ for all $n_c/N \geq 0.19$, and for $n_c/N < 0.19$, the gap $\Delta$ remains nonzero at $T = 0$. For $d > 2$, $\Delta$ vanishes above some critical value of $n_c/N$ for all $T < T_{\rm Néel}(n_c/N)$, the Néel-ordering temperature. Cases where $\Delta = 0$ require $\langle b \rangle$, $\langle \bar{b} \rangle$ to be nonzero due to condensation into the zero-energy states, which is identified physically as long-range Néel order.[9] In this paper, we shall be interested in the *disordered* state at $T = 0$ and $d = 1, 2$ ($n_c/N < 0.19$ for $d = 2$) where $SU(N)$ symmetry is *unbroken*.

When $\Delta \ll J$, the long-wavelength $b$, $\bar{b}$ excitations have a realtivistic spectrum with speed of "light" (spin-wave velocity) $c \sim \bar{\lambda}a/d^{1/2}$ and mass $\Delta/c^2$. The ground state of $H_{\rm MF}$ has the form for $\Delta > 0$

$$|\Omega\rangle \propto \exp\left(\sum_k f_\mathbf{k} g^\dagger_{k a} \bar{b}^{a\dagger}_{-\mathbf{k}}\right)|0\rangle, \quad (2)$$

which represents a condensate of singlet pairs of bosons ("valence bonds"); the bonds have ends on opposite sublattices and their characteristic size is $c/\Delta$. When projected onto $n_c$ bosons per site, $|\Omega\rangle$ is an $SU(N)$ generalization of the short-range resonating-valence-bond states of Sutherland[10] and Liang, Doucot, and Anderson,[10] which are thus *exact* in the present large-$N$ limit provided the distribution of bond lengths is chosen correctly. The eigenmodes of $H_{\rm MF}$ are clearly *bosons* in agreement with recent calculations.[11]

We now consider the fate of the $U(1)$ gauge invariance of $\mathcal{L}$ in the mean-field theory of the disordered state. It is useful to examine first *global* (site and $\tau$ independent) transformations; since our system has two sites per unit cell, there are two such invariances: (i) uniform, $b \to e_i^\phi b$, $\bar{b} \to e_i^\phi \bar{b}$; and (ii) staggered, $b \to e_i^\phi b$, $\bar{b} \to e_{-i}^\phi \bar{b}$. Clearly the "uniform" symmetry is broken by the nonzero value of $\bar{Q} \sim \langle b^\alpha \bar{b}_\alpha \rangle$ while the "staggered" symmetry is not. Considering the full group of *local* gauge transformations we see that it splits into two parts: the uniform part which is broken, and the staggered part which is ot. Fluctuations of $Q$ and $\lambda$ can be written in the form (for each unit cell labeled by $i \in A$)

$$Q_{i,i+\hat{\eta}} = \left[\bar{Q} + q_{\hat{\eta}}\left(i + \frac{1}{2}\hat{\eta}\right)\right] \exp\left[i\theta_{\hat{\eta}}\left(i + \frac{1}{2}\hat{\eta}\right)\right],$$

$$i\lambda(i) = \bar{\lambda} + i\lambda_1(i), \quad i\lambda(i + \hat{x}) = \bar{\lambda} + i\lambda_2(i + \hat{x}),$$

and in momentum space,

$$aA_{\hat{\eta}}(\mathbf{k}) = \frac{1}{2}[\theta_{\hat{\eta}}(\mathbf{k}) - \theta_{-\hat{\eta}}(\mathbf{k})] = -aA_{-\hat{\eta}}(\mathbf{k}),$$

# Valence-Bond and Spin-Peierls Ground States

$$M_{\hat{\eta}}(\mathbf{k}) = \frac{1}{2}[\theta_{\hat{\eta}}(\mathbf{k}) + \theta_{-\hat{\eta}}(\mathbf{k})] = M_{-\hat{\eta}}(\mathbf{k}),$$

$$A_\tau(\mathbf{k}) = \frac{1}{2}[\lambda_1(\mathbf{k}) - \lambda_2(\mathbf{k})],$$

$$M_\tau(\mathbf{k}) = \frac{1}{2}[\lambda_1(\mathbf{k}) + \lambda_2(\mathbf{k})].$$

With $\hat{\eta}$ in a positive axis direction, the $A_{\hat{\eta}}$, $A_\tau$ are the components $(\mathbf{A}, A_\tau)$ of the gauge field for the unbroken, staggered $U(1)$ symmetry, while the $M$'s are related to the broken uniform symmetry. Note that the two modes of $H_{MF}$ at each point $\mathbf{k}$ in the Brillouin zone have charges $\pm 1$ with respect to the staggered symmetry; i.e., they are particle and antiparticle.

We now give the form of the long-wavelength ($\gg a$) effective action of $H$ in terms of the continuum fields $q_{\hat{\eta}}$, $A$, $M$, $z^\alpha \cdot (b^\alpha + \bar{b}_\alpha{}^\dagger)/2$, $\pi^\alpha = (b^\alpha - \bar{b}^{\alpha\dagger})/2$, obtained after integrating out $\pi$:

$$S_{\text{eff}} = \int d^dr \int_0^{c\beta} d\tilde{\tau}\left[\frac{a^{1-d}}{2\sqrt{d}}\left\{|(\partial_\mu - iA_\mu)z^\alpha| + \frac{\Delta^2}{c^2}|z^\alpha|^2\right\}\right.$$
$$\left. + \frac{N}{4e^2}F_{\mu\nu}^2 + iN\gamma\sum_{\hat{\eta}>0}(q_{\hat{\mu}} - q_{-\hat{\mu}})F_{\hat{\mu}\tilde{\tau}}\right]$$

plus additional terms involving $M$ and $q_{\hat{n}}$. Here $\tilde{\tau} = c\tau$, $A_{\hat{\tau}} = A_\tau/c$, and $F_{\mu\nu} = \partial_\mu A_\nu - \partial_\nu A_\mu$, where $\mu$, $\nu$ run over $x, y, \ldots, \tilde{\tau}$, is the electromagnetic field. The terms involving $z$ come from $\mathcal{L}$ while the remaining terms come from integrating out $b$, $\bar{b}$ (or $z$) at one-loop order, giving the coefficient $N$. For $d < 3$, $e^2 \sim (\Delta/c)^{3-d}$ can be calculated in the continuum limit, but $\gamma$ (needed for the spin-Peierls calculation below) has to be calculated using the underlying lattice regularization, giving $\gamma \sim a^{1-d}/\lambda$. The $z$, $A$ part of $S_{\text{eff}}$ has just the form that would be expected by first passing to the continuum semiclassical limit of the Néel phase of $H$ (Ref. 12) and then taking the large-$N$ limit of the resulting $CP^{N-1}$ model.[13]

So far $S_{\text{eff}}$ contains only terms for small fluctuations, but no terms relating to topologically nontrivial gauge-field configurations. These terms, which are expected to be Berry phase factors in the functional integral, would be obtainable by integrating out $b$, $\bar{b}$ in the presence of a nontrivial background gauge field. This should be equivalent to our procedure below of calculating the phase due to adiabatic evolution of the ground state (2) in such a background. We discuss $d = 1, 2$ in turn:

(i) $d = 1$.—The only relevant term which could be added to $S_{\text{eff}}$ is $(i\Theta/2\pi)\int dx d\tilde{\tau}F_{x\tilde{\tau}}$, as suggested by semiclassical calculations[12] in the Néel-ordered phase which produce just this term, when written in $CP^{N-1}$ language,[13] with $\Theta = \pi n_c$. This term survives destruction of Néel order and can be derived *directly* in the disordered phase as follows. Consider a spin chain with $N_s$ sites ($N_s$ even) and periodic boundary conditions. Choosing the configuration in the phase of $Q$, $\theta_{\hat{\eta}}\left(i + \frac{1}{2}\hat{\eta}, \tau\right) = \text{sgn}(\hat{\eta})\phi(\tau)$, where $\phi(\tau)$ increases slowly from 0 at $\tau = 0$ to the gauge equivalent value $2\pi l/N_s$ at $\tau = \beta$ ($l$ integer), yields $\int dx d\bar{\tau} F_{x\bar{\tau}} = 2\pi l$. At $\tau = 0$ we have the wave function $|\Omega\rangle$ in Eq. (2) with $f_k$ real and the sum is over $k = 2\pi n/aN_s$, $n = 1, \ldots, N_s/2$. For $\tau > 0$ we find $\langle \Omega | d/d\tau | \Omega \rangle = 0$; as a result

$$\left|\Omega(\tau = \beta)\right\rangle \propto \exp\left(\sum_k f_{k - 2\pi l/aN_s} b^\dagger_{k\alpha} \bar{b}^{\alpha\dagger}_{-k}\right)|0\rangle. \quad (3)$$

The gauge-invariant Berry phase is now just the change in the phase of the wave function, which for large $N$ is $P_{n_c}|\Omega(\tau = \beta)\rangle = (-1)^{n_c l} P_{n_c}|\Omega(\tau = 0)\rangle$, where $P_{n_c}$ projects onto $n_c$ bosons per site. This phase may be included in $S_{\text{eff}}$ by using $\Theta = p\pi$, where $(-1)_p = (-1)^{n_c}$. Each choice of $\Theta$ corresponds to a different metastable state of the spin chain with a mean static electric field[13] $iF_{x\bar{\tau}} = e^2 p/N$, energy per site $\sim ce^2 p^2 a/N$, and a spin-Peierls order parameter

$$\langle \hat{S}(i) \cdot \hat{S}(i + 1) - \hat{S}(i) \cdot \hat{S}(i - 1)\rangle \sim N\bar{Q}\langle q_{\hat{x}} - q_{-\hat{x}}\rangle/J$$
$$\sim \gamma e^2 cp.$$

The ground state for $n_c$ even is therefore obtained with the choice $p = 0$ and is nondegenerate; the linear Coulomb force confines the spinons in pairs. For $n_c$ odd the ground state corresponds to $p = \pm 1$, and is *twofold degenerate with a nonzero spin-Peierls order parameter;* the spinons are domain walls interpolating between the two ground states. A schematic of the two ground states is shown in Figs. 2(a) and 2(b). The spin-Peierls order for $n_c$ odd was anticipated by Affleck[14] though not shown directly for $n_c \sim N$. This picture is now expected to be correct for *all* $N > 2$.[4,14]

(ii) $d = 2$.—In the Néel-ordered state of the $CP^{N-1}$ model, the Berry-phase term vanishes for any spin configuration which is smooth on the scale of the lattice spacing,[15] but is nonzero for space-time "hedgehog" singularities.[3] In the disordered phase, we use the correspondence between the electromagnetic field tensor $F_{\mu\nu}$ and the "topological charge" $i(\partial_\mu z^*_\alpha \partial_\nu z^\alpha - \partial_\nu z^*_\alpha \partial_\mu z^\alpha)$ of the $CP^{N-1}$ model[13] to identify pointlike instanton configurations of the $(2 + 1)$-dimensional compact $U(1)$ gauge theory[16,17] which have $\int F_{\mu\nu} dS_{\mu\nu} = 2\pi m$ (the integral is over a sphere surrounding the singular point and $n$ is an integer) as the remnants of the hedgehog of the Néel phase. The Berry phase of the instantons can be calculated in a manner very similar to that employed for $d = 1$: We obtain a result (specified below) *identical* to the hedgehog Berry phase calculated by Haldane[3] and its extension to $SU(N)$.[4]

The subsequent analysis follows closely Polyakov's solution[16] of $(2 + 1)$-dimensional compact QED. Neglecting all fields except $A$ at distances $>c/\Delta$, the action is evaluated for each instanton configuration, to give the partition function

## Valence-Bond and Spin-Peierls Ground States

$$Z = \sum_{K,\{m_s\}} \frac{1}{K!} \prod_{s=1}^{K} \left( \sum_{R_a} \int_0^{c\beta} \frac{d\tilde{\tau}_s}{\rho a} \right) \exp[-S_m(\{m_s\})],$$

$$S_m(\{m_s\}) = \frac{N\pi}{2e^2} \sum_{s \neq t} \frac{m_s m_t}{[(\mathbf{R}_s - \mathbf{R}_t)^2} \quad (4)$$

$$+ (\tilde{\tau}_s - \tilde{\tau}_t)^2]^{1/2} + \sum_s \left( NE_c n_s^2 + i\frac{n_c \pi}{2} \zeta_s m_s \right).$$

Note the following: (i) The instantons are represented by integer charges $m_s$ located at $\mathbf{R}_s$, the centers of the plaquettes. (ii) $\rho$ is a dimensionless constant of order unity. (iii) The $1/r$ interaction between instantons is valid at distances larger than the spin-correlation length $c/\Delta$ in contrast to the linear $r$ interaction between hedgehogs on the ordered side. (iv) $NC_c$, the instanton core-action, is determined by the physics at length scales shorter than $c/\Delta$; assuming that the instanton is better described as a hedgehog at these length scales, we expect $E_c \sim \bar{\lambda}/\Delta$. (v) The term proportional to $\zeta_s$ is the Berry phase of the instanton; we have $\zeta_s = 0$, 1, 2, 3 for $\mathbf{R}_s$ on sublattices $W, X, Y, Z$ [Fig. 2(c)]. The well-known equivalence between the $d$-dimensional Coulomb gas and the sine-Gordon model[16] can now be used to show that the long-distance properties of $Z$ are equivalent to those of $Z = \int D\chi \exp(-S_{sG})$ with

$$S_{sG} = \frac{g}{2} \int_0^{c\beta} d\tilde{\tau} \left\{ \sum_{\langle s,t \rangle} (\chi_s - \chi_t)^2 + \sum_s \{a^2(\partial_{\tilde{\tau}}\chi_s)^2 - M^2 \cos\chi_s - (n_c\pi/2)\zeta_s]\} \right\}. \quad (5)$$

Here $\chi$ is the sine-Gordon field which was coupled to the instanton charge with the term $\exp(i\chi_s m_s)$, $g = e^2/4N\pi^2$, and $M^2 = (2/g\rho a)\exp(-NE_c)$ is the *exponentially* small instanton fugacity. In the transformation from Eq. (4) to Eq. (5) we have made the small-fugacity approximation of neglecting instantons with $|m_s| \geq 2$.

If $n_c = 0 \pmod 4$, $S_{sG}$ is the usual sine-Gordon model. For small $M$, it is solved by expanding perturbatively around a minimum.[16] This gives a "screening length" in the instanton plasma $\sim aM^{-1}$ and confinement of $z$ quanta (spinons) into pairs of size $\sim aM^{-1}$. The fluctuations in $F$ give a collective mode of gap $\sim cM/a$. This closely resembles the properties of the valence-bond-solid states recently introduced for $n_c = 2S = 4$ in an $SU(2)$ model,[18] and gives the full lattice symmetry [Fig. 2(f)].

For $n_c \neq 0 \pmod 4$ the uniform state $\chi_s = \text{const}$ is *unstable*. The rotation symmetry between the four sublattices $W, X, Y, Z$ is therefore *spontaneously broken*. For $n_c = 1 \pmod 4$ one stable minimum of $S_{sG}$ is given to order $M^2$ by $\chi_W = \chi_X = -\pi/4 - M^2/4\sqrt{2}$, $\chi_Y = \chi_Z = -\pi/4 + M^2/4\sqrt{2}$ (there are three other similar minima near $\pi/4$, $3\pi/4$, and $-3\pi/4$). This minimum has a static electric field [Fig. 2(c)]: $iE_3 = iE_4 = 0$, $iE_1 = iE_2 = \pi g M^2/\sqrt{2}a$. The coupling between the electric field and the $q_{\hat{n}}$ field in $S_{\text{eff}}$ now implies an exponentially small (in $N$)

but *nonzero spin-Peierls order* of the type shown in Fig. 2(d) with $\langle q_{\hat{x}} - q_{-\hat{x}} \rangle \sim (\gamma \bar{\lambda} a) \pi g c M^2/\sqrt{2}$. A very similar analysis can be performed for $n_c = 3 \pmod 4$. For $n_c = 2 \pmod 4$ the minima of $S_{sG}$ lead to the static electric fields $iE_2 = iE_3 = -iE_4 = gM^2/4a$ and spin-Peierls order of the type shown in Fig. 2(e).[19] These states with broken lattice symmetry also give confinement of spinons and a massive spinless collective mode but with gap (inverse confinement scale) $\sim cM^2/a$ and $cM^4/a$ for $n_c = 2$ and $1,3 \pmod 4$, respectively. This completes are results.

A similar calculation can be carried out for other bipartite lattices, in particular the honeycomb lattice in $d = 2$. This has coordination number 3 and the periodicity in ground-state properties is then in $n_c \pmod 3$, which is consistent with Ref. 18. Our results also generalize to models were sublattice $A$ has $m$ rows in its Young tableau, requiring that the bosons have $U(m)$ gauge symmetry.[4] The only modification to our results is that in $d = 1$ there are no soliton excitations connecting the ground states for $n_c$ odd.

We thank E. Fradkin, T. K. Ng, P. A. Lee, and R. Shankar for discussions. S. S. was supported in part by the NSF.

# REFERENCES

1. J. G. Bednorz and K. A. Muller, *Z. Phys. B* **64**, 188 (1986); M. K. Wu *et al.*, *Phys. Rev. Lett.* **58**, 908 (1987).
2. P. W. Anderson, *Science* **235**, 1196 (1987).
3. F. D. M. Haldane, *Phys. Rev. Lett.* **61**, 1029 (1988).
4. N. Read and S. Sachdev (to be published).
5. Spinless collective modes have also been considered by D. Rokhsar and S. Kivelson, *Phys. Rev. Lett.* **61**, 2376 (1988).
6. D. P. Arovas and A. Auerbach, *Phys. Rev. B* **38**, 316 (1988); *Phys. Rev. Lett.* **61**, 617 (1988).
7. I. Affleck and J. B. Marston, *Phys. Rev. B* **37**, 3774 (1988).
8. N. Read and D. M. Mewns, *J. Phys. C* **16**, 3273 (1983); N. Read, *J. Phys. C* **18**, 2651 (1985).
9. D. Yoshioka (to be published).
10. B. Sutherland, *Phys. Rev. B* **37**, 3786 (1988); S. Liang, B. Doucot, and P. W. Anderson, *Phys. Rev. Lett.* **61**, 365 (1988).
11. N. Read and B. Chakraborty (to be published); H. Levine and F. D. M. Haldane (to be published).
12. F. D. Haldane, *Phys. Lett.* **93A**, 464 (1983); I. Affleck, *Nucl. Phys.* **B257**, 397 (1985).
13. A. D'Adda, P. Di Vecchia, and M. Luscher, *Nucl. Phys.* **B146**, 63 (1978); E. Witten, *Nucl. Phys.* **B149**, 285 (1979); S. Coleman, *Ann. Phys.* (N.Y.) **101**, 239 (1976).

14. I. Affleck, *Phys. Rev. Lett.* **54**, 966 (1985); (to be published).
15. X. G. Wen and A. Zee, *Phys. Rev. Lett.* **61**, 1025 (1988); E. Fradkin and M. Stone, *Phys. Rev. B* **38**, 7215 (1988); T. Dombre and N. Read, *Phys. Rev. B* **38**, 7181 (1988); R. Shankar and S. Sachdev (unpublished).
16. A. M. Polyakov, *Nucl. Phys.* **B120**, 429 (1977); *Gauge Fields and Strings* (Harwood, New York, 1987).
17. P. B. Weigmann, *Phys. Rev. Lett.* **60**, 821 (1988); (to be published).
18. I. Affleck, T. Kennedy, E. H. Lieb, and H. Tasaki, *Phys. Rev. Lett.* **59**, 799 (1987); D. Arovas, A. Auerbach, and F. D. M. Haldane, *Phys. Rev. Lett.* **60**, 531 (1988).
19. In this case the electric field oscillates in sign and we must include both fields $A$, $M$ with $E_{\hat{\eta}}\left(i \pm \frac{1}{2}\hat{\eta}\right) = \pm \partial_\tau \theta_{\pm\hat{\eta}}\left(i \pm \frac{1}{2}\hat{\eta}\right)$ for $i \in A$ and $\hat{\eta} > 0$ in the gauge $\lambda = $ const. The coupling $i(q_{\hat{x}} + q_{-\hat{x}})[\gamma_1(2M_\tau - \partial_\tau M x) + \gamma_2(2M_\tau - \partial_\tau M y)]$ $+ (x \leftrightarrow y)$ in $S_{\text{eff}}$ then leads to the spin-Peierls order shown in Fig. 2(e).

S. C. Zhang
Institute for Theoretical Physics
University of California
Santa Barbara, California 93106

and

T. H. Hansson and S. Kivelson
Physics Department
State University of New York at Stony Brook
Stony Brook, New York 11794

# The Landau-Ginzburg Theory for the Fractional Quantum Hall Effect*

Starting directly from the microscopic Hamiltonian, we derive an effective field theory model for the fractional quantum Hall effect. By considering a coarse grained version of the same model, we construct an approximate Landau-Ginzburg theory, similar to that of Girvin. The partition function of our model exhibits cusps and the Hall conductance is quantized at filling factors $\nu = \dfrac{1}{2k-1}$ with $k$ an arbitrary integer. At these fractions the ground state of our model is incompressible, and the quasi-particles and quasi-holes have fractional charge and obey fractional statistics. Finally we show that the collective density fluctuations are massive.

The route that led to the understanding of the fractional Quantum Hall effect (FQHE)[1] was quite different from the path taken for superconductivity and superfluidity. The macroscopic theories of London and Landau-Ginzburg, which describe the thermo- and electrodynamical properties of superconductors at a phenomenological level, existed long before the microscopic BCS theory. These

---

*Contribution to the Johns Hopkins Workshop on Field Theories in Condensed Matter Physics.

theories have proven to be useful—even after the celebrated success of the microscopic BCS theory—since they can describe long distance phenomena which are only implicitly contained in the BCS theory. Examples are the vortex structure of type II superconductors, fluctuation effects in the phase transition, Josephson junction arrays, *etc*. In the case of FQHE, the situation is just the reverse; the variational approach based on Laughlin's wave function[2] and the cooperative ring exchange[3,4] approach give excellent understanding at the microscopic level. However, as emphasized by Girvin,[5] it is still important to develop an effective field theory model of the FQHE analogous to the Landau-Ginzburg theory of superconductivity. Such a model could not only give a deeper understanding of this remarkable phenomenon, but might also lead to predictions of novel long distance effects inaccessible to the microscopic theories.

An important step in this direction was taken by Girvin,[5] and Girvin and MacDonald,[6] who proposed a field theory model, containing a complex scalar field $\phi$ coupled to a vector field ($a_0$, $\vec{a}$) with a Chern-Simons action (or a topological mass term). This model exhibits vortex solutions with finite energy and fractional charge which can be identified with Laughlin's quasi-particles and quasi-holes. The amplitude fluctuations of the $\phi$ field are massive and are identified with the density fluctuation modes of the single-mode-approximation.[7,8] There is, however, no explanation for why the Hall conductance is quantized at certain specific fractional values, and in Ref. 6 it is also argued that the phase fluctuation modes remain massless, contrary to the belief (based on the microscopic models) that *all* elementary excitations above the ground state have a finite gap, corresponding to an incompressible quantum liquid. Despite this, the model in Refs. 5 and 6 provides an important step towards a complete macroscopic description. In this work, we derive a related model directly from the microscopic Hamiltonian, and demonstrate that it explains almost all known aspects of FQHE.[7]

The basic idea of our approach is to represent the original interacting two-dimensional electron gas by a bose gas problem with a long ranged gauge interaction. This kind of mapping is only possible in two dimensions, where the statistics can be exactly mimicked by the Bohm-Aharonov phase factors resulting from the interaction of the charged particles with the fictitious magnetic flux tubes bound to them. The problem of the bose gas with long ranged gauge interaction can then be represented by a "Chern-Simons" field theory which is accessible to the mean field treatment.

Our model contains one complex scalar bose field $\phi$, whose modulus is the electron density, which is coupled both to a 2 + 1 dimensional "statistical" gauge field ($a_0$, $\vec{a}$) with a Chern-Simons action and to an external electromagnetic vector potential ($A_0$, $\vec{A}$). The coefficient of the Chern-Simons term is determined by demanding that the elementary quanta of the $\phi$-field obey fermi statistics. The resulting model exhibits cusps in the partition function at densities $n = n_B/(2k - 1)$ (where $n_B$ is the density of states in the lowest Landau level) corresponding to the uniform solutions. Moving away from the odd integer filling fractions, the homogeneous state acquires an energy $\sim B|\delta n|$. In fact, the lowest lying charged excitations involve a non-uniform charge density; they are spatially localized vortices with the same charge and statistics as the quasi-particles in Laughlin's approach.[2,9,10] The quasi-particle has a finite creation energy which implies that the system is incompressible. Applying the same logic to a gas of these quasi-particles leads naturally to a hierarchy scheme[9,11] describing the hall conductance fractions of $\frac{2}{5}, \frac{2}{7}$, etc. Finally we also show that the amplitude fluctuation of the φ field has a gap and can be identified with the collective density fluctuations,[8] while the phase fluctuation of the φ field, or the Goldstone boson, is "eaten" by the vector field $(a_0, \vec{a})$ due to the Anderson-Higgs mechanism and disappears entirely from the spectrum.

We start from the following second quantized many body Hamiltonian

$$H = \int d^2 r \phi^*(\vec{r}) \left[ -\frac{1}{2m} (-i\vec{\nabla} - e\vec{A}(\vec{r}) - e\vec{a}(\vec{r}))^2 + eA^0(\vec{r}) \right] \phi(\vec{r})$$

$$+ \frac{1}{2} \int d^2 r d^2 r' \phi^*(\vec{r}) \phi^*(\vec{r}') V(\vec{r} - \vec{r}') \phi(\vec{r}) \phi(\vec{r}') \quad (1)$$

where

$$a^i(\vec{r}) = \frac{\theta}{\pi e} \epsilon^{ij} \int d^2 r' \frac{r_j - r'_j}{|\vec{r} - \vec{r}'|^2} \phi^*(\vec{r}') \phi(\vec{r}') \quad (2)$$

and where we set $\hbar = c = 1$. This Hamiltonian describes a system of identical particles with mass $m$ which are created by the complex scalar field operator φ. These particles interact via a two-dimensional gauge potential $\vec{a}$, and a two body potential $V$. They are also coupled to an external electromagnetic field $A^\mu$. From (2) it is clear that $\vec{a}$ is nothing but the "statistical" gauge potential employed in.[12,13] Thus, if we take the φ field to be bosonic, the above Hamiltonian describes "anyons" obeying θ statistics, (i.e. the wave function changes by a phase θ under interchange of particles). For $\theta = (2k - 1)\pi$ with $k$ an arbitrary integer, this is simply the Hamiltonian for spin polarized electrons in an external electromagnetic field interacting via the two-body potential $V(\vec{r})$.

From (2) we immediately get the following expression for the "statistical" gauge field $b$

$$b(\vec{r}) = -\epsilon^{ij} \partial_i a_j(\vec{r}) = \frac{2\theta}{e} |\phi(\vec{r})|^2 \equiv s|\phi(\vec{r})|^2 \left( \frac{2\pi}{e} \right), \quad (3)$$

which corresponds to associating $\theta/\pi = s$ units of flux to each particle. Changing between different $k$'s corresponds to a singular gauge transformation of the form

$$\delta a_i(\vec{r}) = \frac{2\Delta k}{e} \partial_i \int d\vec{r}' \alpha(\vec{r}, \vec{r}') |\phi(r')|^2 \quad (4)$$

where $\alpha(\vec{r}, \vec{r}')$ is the azimuthal angle of the vector $\vec{r} - \vec{r}'$, which changes the value of $b$ in even integer steps of $(2\pi/e)n$ where $n$ is the density. Although in general singular gauge transformations change the state these specific ones do not.

We can incorporate the constraint (3) by means of a Langrange multiplier field $a_0$, and we add a chemical potential $\mu$, which leads to the following coherent state path integral representation for the partition function

$$Z[A^\mu] = \int [d\phi][da_i^T][da_0] e^{iS[\phi, a_i^T, a_0]} \tag{5}$$

where $a_i^T$ is a transverse gauge field (*i.e.* satisfying $\partial^i a_i^T = 0$), and $S = \int dt d\vec{r} \mathcal{L}$ with

$$\mathcal{L} = i\phi^* \partial_0 \phi - H(\phi) + \mu \phi^* \phi - a_0 \left( \frac{e^2}{2\theta} \epsilon^{ij} \partial_i a_j^T + e\phi^* \phi \right) \tag{6}$$

The term $a_0 \epsilon^{ij} \partial_i a_j^T$ is one half of the famous Chern-Simons term[14,15,16] in the transverse gauge, since

$$\epsilon^{\mu\nu\sigma} a_\mu \partial_\nu a_\sigma \equiv \epsilon^{ij} a_0 \partial_i a_j + \epsilon^{ij} a_i \partial_j a_0 - \epsilon^{ij} a_i \partial_0 a_j \tag{7}$$

The last term vanishes in the transverse gauge, while the second term is equal to the first one upon partial integration. We want to emphasize this rather interesting result. Several authors have pointed out that the excitations of two-dimensional field theories with Chern-Simons terms exhibit fractional statistics, and our derivation starting from the anyon formulation of Wilczek, clearly demonstrates this. Also note that the size of the topological mass term is the one obtained both in our previous analysis of topologically massive 2 + 1 dimensional QED,[17] and in the work of Semenoff.[18] ** So far we have made no approximations, other than those involving the intrinsic ambiguities of the coherent state path integral itself.

In order to apply mean field theory, we have to integrate out the short distance fluctuations of the $\phi(\vec{r})$ field and end up with an effective action which describes the physics at distance scales larger than the magnetic length. This effective action contains a renormalized stiffness constant $\kappa$ replacing the bare mass $1/m$, and an effective interaction strength $\lambda$ replacing the non-local interaction $V(r)$. Since the Chern-Simons term describes the statistics of the particles, which we expect to be the same at all length scales, the $\theta$ parameter should not renormalize. In fact, it has been proven that this is indeed the case in three

---

**We should mention that conflicting claims have been made concerning the statistics of field excitations in the presence of a Chern-Simons term. For a discussion see Ref. 17.

dimensional electrodynamics.[19] The resulting partition function in this approximation is of the form

$$Z[A_\mu] = \int [d\phi da_\mu] e^{i(S_\phi[\Phi,a_\mu,A_\mu]+S_a[a_\mu])} \tag{8}$$

with

$$\mathcal{L}_a = -\frac{e^2}{4\theta} \epsilon^{\mu\nu\sigma} a_\mu \partial_\nu a_\sigma \tag{9}$$

$$\mathcal{L}_\phi = \phi^*[i\partial_0 - e(A_0 + a_0)]\phi$$

$$- \frac{\kappa}{2} \phi^*[-i\vec{\Delta} - e(\vec{A} + \vec{a})]^2 \phi + \mu|\phi|^2 - \lambda|\phi|^4 \tag{10}$$

with $\mu = 2\lambda n$ where $n$ is the density.

This action is similar to the one introduced by Girvin.[5] There are some differences in that we have a $\lambda\phi^4$ term for the scalar field, and we have also incorporated time dependence in the action. The essential difference, however, is that Girvin's Chern-Simons action is for the sum of both the statistical gauge field, and the real external electromagnetic field, while in our case, only the statistical gauge field appears in the Chern-Simons term. In Girvin's case, the partition function corresponding to (5) is independent of the external electromagnetic field, and therefore cannot be used to derive the fractional quantum Hall conductance. Furthermore, the vortex does not carry charge with respect to the real $U(1)$ electromagnetic gauge group. In deriving (10) we have assumed that $V(r)$ is short-ranged; if it is long-ranged the expression can be generalized to include a renormalized interaction $\tilde{V}(r)$ which will be equal to $V(r)$ at long distances. In the spirit of the conventional Landau-Ginzburg theory, we ignore terms with higher powers of $\phi$ and higher derivatives that are generated by the coarse graining procedure, and we treat $\kappa$ and $\lambda$ as phenomenological parameters. This completes the derivation of our model; we now proceed to demonstrate that this effective field theory correctly describes the phenomena related to the FQHE.

First consider the case $A_0 = 0$ and $\epsilon^{ij}\partial_i A_j = -B = $ constant. It is immediately clear that $S$ will be minimized by the trivial constant solution $\phi = \sqrt{n}$, $\vec{a} = -\vec{A}$, $a_0 = 0$. Since the statistical gauge field is related to the density via (3), this solution exists only for $\nu = n/n_B = \pi/\theta = 1/(2k - 1)$. This does not necessarily mean that there is a solution only for the particular fraction corresponding to the $\theta$ chosen in the Lagrangian, since the different choices of $k$ are connected via singular gauge transformations which induce an $\vec{r}$ dependent phase in $\phi$. All these solutions should yield uniform density and have the same energy. However, choosing $k$ so that $\phi$ is constant will certainly optimize the case for a mean field approximation since the classical short distance fluctuations have been eliminated.

To calculate the Hall conductance, we apply an external scalar potential $A_0$ with $\partial_i A_0 = -E_i$, in addition to the vector potential $\epsilon^{ij}\partial_i A_j = -B$. The observable (gauge invariant) current is given by

$$j^i = \frac{\delta S}{\delta A_i} = -\frac{\delta S_\phi}{\delta a_i} = +\frac{\delta S_a}{\delta a_i} = \frac{e^2}{2\theta}\epsilon^{ij}(\partial_0 a_j - \partial_j a_0) \qquad (11)$$

where we used the equation of motion $\delta S/\delta a_i = 0$. In the presence of the external scalar potential, the solution to the equations of motion is $\phi = \sqrt{n}$, $\vec{a} = -\vec{A}$ and $a_0 = -A_0$, thus from (11) the induced current is of the form

$$j^i = \frac{e^2}{2\theta}\epsilon^{ij}E^j = \sigma_H^{ij}E^j \qquad (12)$$

since $\theta = (2k - 1)\pi$, this demonstrates that the Hall conductance is quantized in odd fractions of $e^2/h$.

Let us now analyze what happens when we move away from the odd integer filling fractions. Since the $b$ field is locked at $eb = s2\pi n$ where now $n = n_B + \delta n$, the particles will feel a net field $e\delta b = e(b - B) = s2\pi\delta n$. Each particle (or hole) will acquire a cyclotron energy $= \kappa e\delta b/2$, which implies an energy density $\mathcal{E}$

$$\mathcal{E} = (2k - 1)\pi e B|\delta n|\kappa \qquad (13)$$

For large $B$ it is natural to assume that by moving away from the good filling fractions one creates localized density disturbances. As we shall see, our model exhibits such quasi-particle and quasi-hole excitations in the form of vortices similar to those found in Refs. 2, 3, and 5.

From the equation of motion derived from Eq. (8)–(10) one easily finds that at $n = \pi n_B/\theta$ there are static, nonuniform, finite energy vortex solutions. If $(r, \varphi)$ are polar coordinates with the center of the vortex at $r = 0$, the $r = \infty$ behavior of the solution is

$$\phi(r, \varphi) = \sqrt{n}\, e^{\pm i\varphi} \qquad (14)$$

$$\vec{a}(r, \varphi) = \pm\frac{\hat{\varphi}}{er} \qquad (15)$$

and $a_0(r, \varphi) = 0$, corresponding to one unit of statistical flux per vortex. The equation of motion for $a^0$ implies

$$j^0 \equiv -\frac{\delta S}{\delta A_0} = -\frac{\delta S_\phi}{\delta a_0} = \frac{\delta S_a}{\delta a_0} = \frac{e^2}{2\theta}b \qquad (16)$$

so the total charge carried by the vortex

$$q_1 = \int d^2x j^0 = \frac{e^2}{2\theta} \oint \vec{a} \cdot \vec{dr} = \pm\frac{\pi e}{\theta} = \pm\frac{e}{2k-1} \qquad (17)$$

These field configurations can thus be identified as the fractionally charged quasi-particle and quasi-holes above the ground state. According to the results of Refs. 12, 10, 13 and 18, 19 this implies that the quasi-particles obey fractional statistics with $\theta_1 = q_1\Phi_1/2 = \pi/(2k-1)$, where $\Phi_1 = 2\pi/e$ is the flux of the vortex. Note that since $\phi(\vec{r})$ must vanish at the center of the vortex, there is necessarily a difference in the profile and creation energies of the quasi-particles and quasi-holes. (This also illustrates the intrinsic problems with the Landau-Ginzburg approach at distances of order the magnetic length, since the true quasi-particle density certainly does not vanish in the core.)

We have thus shown that the vortices in our model have the same charge and statistics as the quasi-particles in Laughlin's approach.[2,9,10] The presence of these excitations naturally leads to the so-called hierarchy scheme[9,10] which has been proposed to explain the quantization of the Hall conductance at fractions other than $1/(2k-1)$. Let $S(q_s, \theta_s)$ be the action in (7)–(9) with charge $q_s$ and $\theta$-parameter $\theta_s$; $S(q_s, \theta_s)$ describes the physics at the $s^{th}$ level of the hierarchy. At the zeroth level, $s = 0$, we have $q_0 = e$ and $\theta_0 = (2k-1)\pi$ which describes a system of electrons and we have shown that the filling factor $\nu$ is pinned at $\nu_0 = \pi/\theta_0$ and the Hall conductance is given by $\sigma_0 = q_0^2/(2\theta_0)$. If the filling factor deviates from the special values corresponding to level $s$, extra quasi-particles or quasi-holes have charge $\pm(\pi/\theta_s)q_s$ and a statistical angle $\pi^2/\theta_s$. A system of these extra quasi-particles or quasi-holes can therefore be described by a new action $S(q_{s+1}, \theta_{s+1})$ with

$$q_{s+1} = \pm q_s \frac{\pi}{\theta_s} \quad \text{and} \quad \theta_{s+1} = \pi\left(2k \mp \frac{\pi}{\theta_s}\right) \qquad (18)$$

The Hall conductance is given by $\sigma_{s+1} = \sigma_s \pm q_{s+1}^2/(2\theta_{s+1})$. In these equations, the upper sign refers to the quasi-particles and the lower sign refers to the quasi-holes. These recursion relations are identical to those derived from the Laughlin's wave function approach[9,11] and describe the Hall conductance at fractions like $\frac{2}{5}, \frac{2}{7}, \frac{4}{11}, \frac{4}{13}$, etc.

Finally let us turn to the collective excitations. Due to the symmetry breaking potential in $L_\phi$, the $\phi$ field acquires a non-vanishing vacuum expectation value $|\langle\phi\rangle| = \sqrt{n}$. The $\phi$ field can thus be parametrized by $\phi(x) = (\phi_0 + \delta\phi(x))e^{i\eta(x)}$, $\vec{a}(x) = \delta\vec{a}(x) + \vec{\nabla}\eta(x)$, and $a_0(x) = \delta a_0(x) - \partial_0\eta(x)$, describing the amplitude and phase fluctuations about the classical vacuum. We see that the phase fluctuation $\eta(x)$ is "gauged away" in accordance with the standard Anderson-Higgs mechanism. Since the statistical gauge field is non-dynamical, there is no propagating mode (massive or not) corresponding to phase fluctuations. This reflects that there is a unique ground-state in the quantum Hall effect and, we believe, implies that there will be no Josephson-like effects. Only the amplitude fluctuation remains, and by expanding the Lagrangian, up to terms quadratic in $\delta\phi$, $\delta\tilde{a}$, and $\delta a_0$, about the constant solution, we find the following dispersion relation

$$\omega(q)^2 = (e\kappa B)^2 + \frac{\kappa}{4} q^2(\kappa q^2 + 8\lambda\phi_0^2). \tag{19}$$

Note that the mass of the amplitude mode is $\sim B$, and for negative $\lambda$ the dispersion curve has the same shape as that derived in Ref. 8. Note that even for negative $\lambda$, as long as $|\lambda|/\kappa$ is sufficiently small, the quasi-particle creation energy is positive, and the Hamiltonian is bounded from below.

In conclusion, we have derived the Landau-Ginzburg theory for the FQHE directly from the microscopic Hamiltonian, and find that this effective field theory describes almost all the known phenomenology including incompressibility, fractional Hall conductance with odd denominators and the fractional charge and statistics of the quasi-particles. It is, however, to be warned that this effective theory certainly makes errors on the magnetic length scale, and it treats the statistical gauge field $\tilde{a}$ within mean field theory, *i.e.* the particles feel the $b$ field which produces the statistics, whereas in fact the exact $\tilde{a}$ is pure gauge. Despite these shortcomings, we believe that the long-wavelength properties of the quantum Hall system are correctly reproduced by this Landau-Ginzburg theory. We note that the same physical idea, *i.e.* that the long-wave-length effects of the physical magnetic field are cancelled by the statistical field, is the basic result of the cooperative-ring-exchange theory[3,4] of the quantum Hall effect, and of a mean-field theory recently introduced by Laughlin.[20] We believe that the success of our "two-dimensional bosonization" approach also suggests the potential applications in treating another strongly correlated two dimensional fermion problem such as the Mott-Hubbard insulators, just like the one-dimensional bosonization approach has proven to be extremely useful to understand the complex phases of the one dimensional electron gas and spin chains. We are also currently engaged in trying to understand the finite temperature effect and the localization problem of the FQHE from this point of view.

While our approach starts directly from the microscopic Hamiltonian, in an alternative approach, N. Reed[21] has derived a similar linearized effective action from the Laughlin's wave function.

# ACKNOWLEDGEMENTS

We would like to thank V. Emery, S. Girvin, D. H. Lee, N. Reed and X. G. Wen for encouraging discussions.

## REFERENCES

1. R. E. Prange and S. M. Girvin, The Quantum Hall Effect, Springer-Verlag (1986).
2. R. B. Laughlin, *Phys. Rev. Lett.* **50**, 1395 (1983).
3. S. Kivelson, C. Kallin, D. P. Arovas and J. R. Schrieffer, *Phys. Rev. Lett.* **56**, 873 (1986).
4. D. H. Lee, G. Basharan and S. Kivelson, *Phys. Rev. Lett.* **59**, 2467 (1987).
5. S. M. Girvin, in chap. 10 of Ref. 1.
6. S. M. Girvin and A. H. MacDonald, *Phys. Rev. Lett.* **58**, 1252 (1987).
7. S. C. Zhang, H. Hansson and S. Kivelson, *Phys. Rev. Lett.* **62**, 82 (1989).
8. S. M. Girvin, A. H. MacDonald and P. M. Platzman, *Phys. Rev. Lett.* **54**, 581 (1985); S. M. Girvin, A. H. MacDonald and P. M. Platzman, *Phys. Rev. B* **33**, 2481 (1986).
9. B. I. Halperin, *Phys. Rev. Lett.* **52**, 1583 (1984).
10. D. P. Arovas, J. R. Schrieffer and F. Wilczek, *Phys. Rev. Lett.* **53**, 722 (1984).
11. F. D. M. Haldane, *Phys. Rev. Lett.* **51**, 605 (1983).
12. F. Wilczek, *Phys. Rev. Lett.* **49**, 957 (1982).
13. D. P. Arovas, J. R. Schrieffer, F. Wilczek and A. Zee, *Nucl. Phys.* **B251**, 117 (1985).
14. W. Siegel, *Nucl. Phys.* **B156**, 135 (1979).
15. J. F. Schonfeld, *Nucl. Phys.* **B185**, 157 (1981).
16. S. Deser, R. Jackiw and S. Templeton, *Ann. Phys.* **140**, 372 (1982).
17. T. H. Hansson, M. Roček, I. Zahed and S. C. Zhang, *Spin and statistics in massive* $2 + 1$ *dim. QED*, preprint, State University of New York at Stony Brook, ITP-SB-88-32 (1988).
18. G. W. Semenoff, *Phys. Rev. Lett.* **61**, 517 (1988).
19. G. W. Semenoff, P. Sodano and Y. S. Wu, preprint.
20. R. B. Laughlin, *Phys. Rev. Lett.* **60**, 2677 (1988).
21. N. Reed, *Phys. Rev. Lett.* **62**, 85 (1989).

**Timothy Ziman**
Department of Physics and Astronomy
University of Delaware
Newark, Delaware 19716

# One-Dimensional Magnets, Wess-Zumino Models and Conformal Invariance

There are a number of insulating materials whose magnetic ions provide a good realization of one-dimensional antiferromagnets of various quantum spin $S$ [1]: $CuCl_2 2NC_5H_5$ which has spin $S = \frac{1}{2}$, $CsNiCl_3$ ($S = 1$) $CsVCl_3$ $\left(S = \frac{3}{2}\right)$, and $(CH_3)_4NC_5MnCl_3$, (also known as TMMC), which has $S = \frac{5}{2}$. Each can, to a first approximation, be described by the very simple Hamiltonian

$$\mathcal{H} = \sum_n \mathbf{S}_n \cdot \mathbf{S}_{n+1} \tag{1}$$

where $\mathbf{S}_n$ is an operator of quantum spin $S$ at site $n$ of a chain. The Hamiltonian is clearly invariant under a global rotation of spins. All of these materials are highly anisotropic with respect to lattice direction as far as magnetic interactions are concerned: the moments form well coupled chains but have weak couplings from chain to chain. The task of the theorist is to calculate thermodynamic functions and dynamical correlation functions which can be measured in some detail by inelastic neutron scattering. Of course much of the interest is more general: we want to use such Hamiltonians as testing grounds for the development of proper theories of strongly interacting many-body systems. If we treated the above spins as classical vectors we would have a ground state with long-range antiferromagnetic order: a Néel-type state with sublattice magnetization in an arbitrarily chosen direction of space. We now know that if the materials were strictly one-dimensional none of them would have such order even at zero temperature. It is well known that order is destroyed by thermal fluctuations at arbitrarily low temperature in one dimension, but here the statement is that zero-point quantum fluctuations are sufficient on their own. This is in contrast to the isotropic ferromagnet where the saturated state is an exact eigenvector. Even though there is no true long range antiferromagnetic order the susceptibility corresponding to antiferromagnetic order may (or may not, depending on the value of the spin) diverge, and there may (or may not) be response at zero temperature and long wavelengths that is at vanishingly low energies. If there is a divergent susceptibility, $T = 0$ can be considered a critical point and we may expect behaviour that is interesting and not necessarily simple.

Having posed the problem and said that it is not necessarily easy, we may ask, what is the proper way to study it? In low dimension perturbation theory and mean-field decoupling schemes are notoriously unreliable. There are two main avenues that have been explored that do give reliable if limited information and the challenge in recent years has been to fit the various pieces of information so derived into a coherent whole. The first is to look for Hamiltonians that can be exactly integrated by the Bethe Ansatz: this approach dates all the way back to the original work by Bethe on the Hamiltonian (1) for $S = \frac{1}{2}$ [2]. More recently Hamiltonians have been solved [3, 4] for higher spin $S$ but the big difference from spin one half is that in order for the Yang-Baxter equations to be satisfied and the model to be integrable, the Hamiltonian has to be rather special looking, and as we shall make precise further on, indeed rather atypical. For example for spin $\frac{3}{2}$ equation (1) is not integrable but the following is:

$$\mathcal{H}_b = \sum_n \left[ -(\mathbf{S}_n \cdot \mathbf{S}_{n+1}) + \frac{8}{27}(\mathbf{S}_n \cdot \mathbf{S}_{n+1})^2 + \frac{16}{27}(\mathbf{S}_n \cdot \mathbf{S}_{n+1})^3 \right] \qquad (2)$$

Evidently Bethe's method provides a solution to models that include more than the simple quadratic exchange that follows from super-exchange. They have extra terms with special coefficients of bi-quadratic and higher (up to order $2S$) powers of the nearest neighbour exchange. It is also important to realize what Bethe solutions actually provide: they give eigenvalues of low-lying states, with some difficulty they can be used to calculate thermodynamics and only rather limited information on correlation functions [2]. So while the role of the Bethe Ansatz is crucial, very important questions are left unanswered: the obvious one being "are the soluble cases special in some way or generic?". Specifically does the ground state of the Hamiltonian (2) have the same correlations as (1) or are they qualitatively different? By qualitatively different, we mean a change in analytic form at long distances and times, not simply a change in spin-wave velocity. For example there may be a change in the exponent characterising the power law decay of some correlation function or exponential decay may develop rather than power-law behaviour. Furthermore, if we cannot calculate all the correlation functions we want, can we at least predict their qualitative form?

The second approach, which is very much in the spirit of the modern school of applying quantum field theoretic methods to many-body problems, is to attempt to map the lattice models onto continuum field theories. One way is to generate a semiclassical or large spin $S$ expansion [5, 6]: this follows Dyson's spin-wave expansion around a spontaneously symmetry-broken state in higher dimension, but treats the resultant continuum theory in a non-perturbative fashion. A more systematic method, though less direct in appearance, starts from the Jordan-Wigner transformation for spin one-half in order to write an equivalent theory of one-dimensional fermions [7]. Higher spin may be built up by taking different species of fermions [8]. This may be treated in the modern fashion of bosonization and the full understanding of two-dimensional critical phenomena brought to bear [9, 10, 11]. A drawback to both the spin-wave and fermionic approaches is that they require us to select out a privileged direction in spin space. If we start with rotationally invariant Hamiltonians such as (1) and (2) this is an unattractive feature, since if the explicit symmetry is not preserved in the formalism it may easily be lost in the course of approximation. This is of course much more serious than in three dimensions where the existence of a spontaneously chosen direction for the ground state provides the choice of spin quantisation direction. There is a way around this known as non-Abelian bosonization: "non-Abelian" in so far as it preserves the (non-Abelian) symmetry $SU(2)$, in contrast to the Jordan-Wigner method which preserves only the Abelian $U(1)$ symmetry of rotations about the fixed axis that is chosen as the direction of quantization. The disadvantage of non-Abelian bosonization is that it cannot be done as explicitly as the Abelian method, ie more careful examination is needed to determine which continuum theory corresponds to the original Hamiltonian. Thanks to work by Witten [12], we know how to recover $SU(2)$ theories that are critical in one spatial and one time dimension: the action must include a topological term with integer coefficient $k$.

$$S = \frac{k}{8\pi} \int d^2x \, \text{tr}(\partial_\mu g \partial^\mu g^{-1}) + \frac{k}{12\pi} \int d^3x \epsilon^{\mu\nu\lambda} \, \text{tr}(g^+\partial_\mu g g^+ \partial_\nu g g^+ \partial_\lambda g) \qquad (3)$$

In this action the first integral is over the two Euclidean dimensions, the three dimensional integral is of fields $g(x) \in SU(2)$ analytically continued from a given two-dimensional configuration of finite action. Because of this continuation the integral is unique only up to multiples of an integer times $24\pi^2$; thus $\exp[iS]$ is only well-defined when $k$ is integer. The realization that the topological term could restore criticality was vital since without it we know that fluctuations would not only destroy long range order but also generate a characteristic length scale: just as a two-dimensional classical Heisenberg model at finite temperature has a finite correlation length. Witten established that for each integer value of $k$ there is a distinct critical point possessing a chiral $SU(2)_L \times SU(2)_R$ symmetry as well as conformal symmetry. Further on, we shall see what this symmetry, higher than the single $SU(2)$ symmetry we started with, implies in terms of spin correlations.

## RÔLE OF CONFORMAL INVARIANCE

As we have said the two approaches cited each suffer limitations: the Bethe Ansatz because in some ways it is simply a machine for churning out certain numbers for certain theories that happen to satisfy integrability conditions, the continuum theory because it really defines possible long wavelength theories rather than that which applies to a given lattice model. It has turned out that the important glue for assembling the slightly fragmented view is provided by the concept of conformal invariance, as developed primarily by Belavin, Polyakov and Zamalodchikov [13], and, for the $SU(2)$ symmetric case, by Knizhnik and Zamalodchikov [14]. We are by now quite familiar with the concept that at a critical point there is no intrinsic length scale for long wavelength fluctuations and therefore the analytic form of correlations is unchanged by a global change of scale. Conformal invariance tells us much more about the correlation functions in the continuum theory. It provides an algorithm: finite-size scaling, for identifying the soluble cases, for example (2), to continuum theories. By giving a more detailed theory of correlation functions it permits a stability analysis for each critical point: this is what allows us to say whether (2), for example, is special or whether slight perturbations correlations may change. Conformal theory does this by providing a detailed theory of finite-size effects that defines a numerical procedure to reliably classify Hamiltonians, gives a theory for finite temperature and allows us to extract much more information from the Bethe Ansatz. By 'numerical' I do not necessarily mean approximate: much of the work can be done from the Bethe equations analytically.

The statement of conformal invariance is an extension of the normal notion of scale invariance: for power law decay

$$\langle \phi(\mathbf{r})\phi(\mathbf{r}') \rangle = A|\mathbf{r} - \mathbf{r}'|^{-p} \qquad (4)$$

enjoys self similarity under the transformation $\mathbf{r} \mapsto \lambda \mathbf{r}$:

$$\langle \phi(\lambda\mathbf{r})\phi(\lambda\mathbf{r}') \rangle = \lambda^{-\frac{p}{2}}\lambda^{-\frac{p}{2}} \langle \phi(\mathbf{r})\phi(\mathbf{r}') \rangle \qquad (5)$$

$$= |f'(\mathbf{r})|^{-\frac{p}{2}}|f'(\mathbf{r}')|^{-\frac{p}{2}} \langle \phi(\mathbf{r})\phi(\mathbf{r}') \rangle$$

where $f(\mathbf{r}) = \lambda \mathbf{r}$; $f'(\mathbf{r}) = \lambda$. It was suggested by Polyakov that provided the critical theory is local, (5) should be satisfied for a transformation that is nonuniform but still locally scale invariant, ie a smooth angle-preserving transformation. In two dimensions this statement of conformal invariance is very powerful because we know that if we take the coordinates to be the real and imaginary parts of a complex number, the conformal group is generated by the infinite dimensional group of complex analytic functions. For the one-dimensional quantum systems we are considering there are two dimensions: space and imaginary time. We can take the two Euclidean variables $x_1 = itv$ and $x_2 = n$, where $v$ is the spin-wave velocity. For finite size scaling the important analytic function, as pointed out by Cardy [15, 16] is the complex logarithm

$$w = \frac{L}{2\pi} \log z \qquad (6)$$

which maps the complex plane to a cylinder of radius $L$. We should think of the finite direction as space and the infinite direction as (complex) time. Then the statement (5) for this transformation is:

$$\langle \phi(w)\phi(w') \rangle = \left|\frac{L}{2\pi r}\right|^{-\frac{p}{2}} \left|\frac{L}{2\pi r'}\right|^{-\frac{p}{2}} |\mathbf{r} - \mathbf{r}'|^{-p} \qquad (7)$$

$$= \left(\frac{2\pi}{l}\right)^{p} \left(\frac{r}{r'}\right)^{-\frac{p}{2}} \left[\frac{|\mathbf{r} - \mathbf{r}'|}{r}\right]^{-p}$$

ie conformal invariance relates the exponential decay on the cylinder to power law decay in the infinite plane. Taking $z' = 1 + i0$, $w' = 0$

$$\langle \phi(w)\phi(0) \rangle = \left(\frac{2\pi}{L}\right)^{p} \exp\left[-\left(\frac{\pi}{L}(\text{Re } w)p\right)\right]\left\{1 + \frac{p}{2}(e^{-\frac{2\pi}{L}w} + e^{-\frac{2\pi}{L}\bar{w}}) + \cdots\right\} \qquad (8)$$

The correlation is an exponential in $w$, $\exp\left[-\frac{\pi}{L}pw\right]$ with corrections differing in the rate of exponential decay by integer multiples of $\frac{2\pi}{L}$. But we can calculate correlations in a finite strip by diagonalising a transfer matrix, which in the case of the quantum spin chains, is simply the exponential of the Hamiltonian we started off with.

$$\langle S_n^z(t) S_n^z(0) \rangle = \sum_m e^{it(E_0 - E_m)} \langle 0|s_n^z|m\rangle\langle m|S_n^z|0\rangle \qquad (9)$$

and the coefficients of the exponent are simply the energies of excitation from the ground state to excited eigenvalues. Identifying the real part of $w$ with $itv$, the leading coefficient of Re $w$ is read off equation (9) as $\frac{E_1 - E_0}{v}$, but comparing to (8) this is equal to $\frac{\pi}{L} p$ where $p$ is the power law describing the decay of correlations in the infinite plane. Clearly there must also be higher levels spaced at $\frac{\pi}{L}(p+2)$, $\frac{\pi}{L}(p+4)$, $\frac{\pi}{L}(p+6)$, ... ie the eigenvalues occur in 'conformal towers', the lowest level in units of $\frac{\pi}{L}$ being the fractional critical exponent the others at integer levels above. For quantum spin chains the rescaling of space and time by means of the velocity $v$ must, in general, be done numerically. Thus the relation reads

$$E_t - E_0 = \frac{v\pi}{L} \eta_t \qquad (10)$$

where $E_t$ is the lowest triplet and $\eta_t$ is the law of power-law decay of the correlation function

$$\langle S_n^z S_0^z \rangle \sim \frac{(-1)^n}{n^{\eta_t}} \qquad (11)$$

This is the first consequence of conformal invariance; there are others. Most important is that from the length dependence of the ground state energy per site $\epsilon(l)$ one can calculate the conformal central charge $c$, the anomalous term in the correlation function of the stress energy tensor.

$$\epsilon(L) = \epsilon_\infty - \frac{\pi}{6} \frac{c}{L^2} v \qquad (12)$$

This number, of central importance in the field theory, becomes an experimental observable at low temperature using the relation between the effective length and temperature [17, 18]. At finite temperature $T$ acts in effect as a cutoff with $L \sim \frac{\hbar v}{k_B T}$. Putting this value in (12) and differentiating it is found that the specific heat is given by

$$C = c \frac{\pi}{3\hbar v} k_B^2 T \qquad (13)$$

In fact this is how the soluble spin chains, such as (2), were first identified to the higher Wess-Zumino-Witten models with $c = \frac{3k}{2+k}$ and $k = 2S$ [18]. Note that the number $c$ is fractional for $k > 1$ and therefore cannot be understood as the number of independent boson excitations as a naive interpretation of (13) would suggest.

When we look at (9) it is clear that the states **m** contributing are triplets under the group of global rotations; if we took different correlation functions contributions would come from different intermediate states **m** with symmetry depending on the operator acting on the vacuum **0**. Therefore we can extract exponents describing the decay of different observables in the infinite plane by looking at excited levels of the finite chain having different quantum numbers. An interesting example of this is the exponent $\eta_s$:

$$\langle (\mathbf{S}(n)\cdot\mathbf{S}(n+1))(\mathbf{S}(1)\cdot\mathbf{S}(0))\rangle \sim \frac{(-1)^n}{n^{\eta_s}} \tag{14}$$

The exponent $\eta_s$ is just as for the $\eta_t$ for the spin-spin correlation function found from a gap but a gap to a low-lying singlet state, since a state **m** such that $\langle \mathbf{m}|\mathbf{S}(1)\cdot\mathbf{S}(0)|0\rangle$ is non-vanishing must be a singlet. For any finite system the lowest excited singlet every $E_s$ is different from the lowest triplet $E_t$ but it is a non-trivial consequence of the chiral $SU(2)\times SU(2)$ symmetry of the Wess-Zumino models that asymptotically these energies must converge and that the exponents $\eta_s$ and $\eta_t$ are the same. The low-lying singlet and triplet levels may be considered a supermultiplet corresponding to $\frac{1}{2}_L \otimes \frac{1}{2}_R$ of the critical theory. The leading corrections to the two effective exponents for a finite length are different but dominated by a single marginal operator, as predicted from the conformal structure. Defining effective exponents from the length dependence of $E_s$ and $E_t$

$$\eta_s(L) = \eta_\infty + 3\frac{b\pi}{2}g(L) + \cdots$$
$$\eta_t(L) = \eta_\infty - \frac{b\pi}{2}g(L) + \cdots \tag{15}$$

where $g(L)$ is a marginal operator, $b$ is the normalised coefficient of a three-point function. More generally one can predict the whole pattern of asymptotically degenerate levels and their relative splittings [14, 19]. If $g(L)$ is extremely small, as can be seen from the actual splittings for length $L$, then $g(L)$ is well approximated by its asymptotic form

$$g(L) = \frac{g}{1+\pi g b \ln\left(\frac{L}{L_0}\right)} \approx \frac{1}{\pi b \ln L} \tag{16}$$

where $b$ is the number $4/\sqrt{3}k$. This asymptotic form will only be applicable in practice either by judicious choice of couplings of the original Hamiltonian or by taking $L$ extremely large for soluble cases. Equation (15) will be good, however, even when this logarithmic régime is not yet reached. The importance of this is that for a given lattice model one may eliminate $g(L)$ from one's estimate of $\eta_\infty$ simply by taking the weighted average $(3\eta_t + \eta_s)/4$.

In terms of the correlation functions in the infinite limit this gives multiplicative logarithmic corrections to the power laws (11) and (14).

$$\langle S_n^z S_0^z \rangle \sim \frac{(-1)^n}{n^{\eta_t}} \log^{\sigma_t}(n) \qquad (17)$$

where $\sigma_t = \frac{1}{2}$ and $\sigma_s$, the analogous correction to (14) is $\sigma_s = -\frac{3}{2}$.

Once all the levels are known and equivalently the scaling dimensions of all operators at a given fixed point one may answer the question of the local stability: if any operator with $\eta \leq 4$ is present then the fixed point will be unstable to perturbation by that operator. This allows one to answer the question of whether going infinitesimally from (2), for example, towards (1) represents a relevant perturbation of the $k = 3$ Witten-Wess-Zumino critical point. In fact it is a relevant perturbation and one may show convincingly using (15) and (12) that crossover is to the $k = 1$ fixed point [20, 21]. Thus the generic behaviour of the spin-$\frac{3}{2}$ antiferromagnet is critical at zero-temperature, but with the exponents of the soluble spin-$\frac{1}{2}$ model, *not* those of the soluble (2). By generic here I mean for a finite region in Hamiltonian space around (1). If one varies the Hamiltonian by a finite amount, for example by adding an isotropic second-nearest neighbour exchange of sufficient strength, there is a transition to a different phase: one with spontaneously dimerized 'spin-Peierls' or 'valence-bond' order. This phase has a spontaneously broken discrete symmetry and a gap to excitations. Such behaviour occurs for all half-integer spins; for integer spins, again the integrable model is critical but unstable with respect to a small perturbation in the direction of (1). In this case the simple exchange (1) is massive with a gap [6, 21].

In conclusion, by use of the concepts of conformal invariance it is possible with a mixture of analytical calculation and numerical analysis to build up a coherent picture of quantum Heisenberg chains. The exactly integrable models, except for spin-$\frac{1}{2}$, are seen to be rather special, as multicritical points in phase space. As one moves from the integrable model towards the simple Heisenberg model (1) it is true in both integer and half-integer case that there is crossover from the higher Witten-Wess-Zumino model. That the crossover is to quite different cases, to the same behaviour as spin $\frac{1}{2}$ for half-integer spin and to massive behaviour for integer spin shows the care with which one must treat semi-classical expansions: the difference between spin $S$ and $S + \frac{1}{2}$ appears by means of different crossover at lengths greater than a divergent correlation length. The fact that the higher spin integrable models do not look physically applicable does not mean that they cannot be relevant to real materials: in fact if we take anisotropic spin models the higher Wess-Zumino models may appear at multicritical points in parameter space. For instance for a spin-1 Hamiltonian

$$\mathcal{H} = \sum_n [J_z S^z_n S^z_{n+1} + (S^x_n S^x_{n+1} + S^y_n S^y_{n+1})] + D \sum_n (S^z_n)^2 \qquad (18)$$

has a multicritical point at $J_z \approx 3$, $D \approx 2.7$ [10, 11] where two singlet phases and the antiferromagnetic phase meet is predicted to have the correlations of the Wess-Zumino model for $k = 2$, even though while it is axially symmetric it is clearly not invariant under general rotations.

## I. ACKNOWLEDGEMENTS

This manuscript was prepared during visits to the Mathematical Sciences Research Institute, Berkeley and the Institute of Theoretical Physics, Santa Barbara, which is supported by the National Science Foundation under Grant PHY 82-17853 supplemented by funds from the National Aeronautics and Space Administration.

## REFERENCES

1. M. Steiner, J. Villain, and C. G. Windsor, *Adv. Phys.* **25**, 87 (1976).
2. R. J. Baxter, "Exactly Solved Models in Statistical Mechanics," Academic Press, London, 1982. This book is an invaluable guide to the Bethe Ansatz and related methods.
3. L. Takhtajan, *Phys. Lett.* **A87**, 479 (1982).
4. H. M. Babudjian, *Phys. Lett.* **A90** (1982); *Nucl. Phys.* **B215** [FS 7], 317 (1983).
5. J. Villain, *J. Phys.* (Paris) **35**, 27 (1974).
6. F. D. M. Haldane, *Phys. Lett.* **93A**, 464 (1983); *Phys. Rev. Lett.* **50**, 1153 (1983).
7. A. Luther and I. Peschel, *Phys. Rev. B* **12**, 3908 (1973).
8. J. Timonen and A. Luther, *J. Phys. C* **18**, 1439 (1985).
9. M. D. M. den Nijs, *Phys. Rev. B* **23**, 6111 (1981); J. L. Black and V. J. Emery, *Phys. Rev. B* **23**, 429 (1981).
10. H. J. Schulz and T. Ziman, *Phys. Rev. B* **33**, 6546 (1986).
11. H. J. Schulz, *Phys. Rev. B* **34**, 6372 (1986).

12. E. Witten, *Commun. Math. Physics* **92**, 455 (1984).
13. A. A. Belavin, A. M. Polyakov and A. B. Zamolodchikov, *Nucl. Phys.* **B241**, 333 (1984).
14. V. G. Knizhnik and Z. B. Zamolodchikov, *Nucl. Phys.* **B247**, 83 (1984).
15. J. L. Cardy, *J. Phys.* **A17**, L385 (1984).
16. J. L. Cardy, *Nucl. Phys.* **B270** [FS 16], 186 (1984).
17. J. L. Cardy, *J. Phys.* **A19**, L1093 (1986).
18. I. Affleck, *Phys. Rev. Lett.* **55**, 1355 (1985), I. Affleck, *Phys. Rev. Lett.* **56**, 746 (1986).
19. I. Affleck, D. Gepner, H. J. Schulz, and T. Ziman, *J. Phys. A: Math. Gen.* **22**, 511 (1989).
20. T. Ziman and H. J. Schulz, *Phys. Rev. Lett.* **59**, 140 (1987).
21. I. Affleck and F. D. M. Haldane, *Phys. Rev. B* **36**, 5291 (1987).